Simon J. Otten

Characterisation of REBCO Roebel cables

Karlsruher Schriftenreihe zur Supraleitung Band 024

HERAUSGEBER

Prof. Dr.-Ing. M. Noe
Prof. Dr. rer. nat. M. Siegel

Eine Übersicht aller bisher in dieser Schriftenreihe erschienenen Bände finden Sie am Ende des Buches.

Characterisation of REBCO Roebel cables

by
Simon J. Otten

KIT Scientific Publishing

Karlsruher Institut für Technologie
Institut für Technische Physik

Characterisation of REBCO Roebel cables

Zur Erlangung des akademischen Grades eines Doktor-Ingenieurs von der KIT-Fakultät für Elektrotechnik und Informationstechnik des Karlsruher Instituts für Technologie (KIT) genehmigte Dissertation

von M.Sc. Simon J. Otten, geb. in Zwolle, die Niederlande

Tag der mündlichen Prüfung: 25. Januar 2019
Hauptreferent: Prof. Dr.-Ing. Mathias Noe
Korreferent: Prof. Dr. Ir. Herman ten Kate

Impressum

KIT Scientific Publishing

Karlsruher Institut für Technologie (KIT)
KIT Scientific Publishing
Straße am Forum 2
D-76131 Karlsruhe

KIT Scientific Publishing is a registered trademark of Karlsruhe Institute of Technology.
Reprint using the book cover is not allowed.

www.ksp.kit.edu

Print on Demand 2019 – Gedruckt auf FSC-zertifiziertem Papier

ISSN 1869-1765
ISBN 978-3-7315-0904-2
DOI 10.5445/KSP/1000091629

Characterisation of REBCO Roebel cables

Zur Erlangung des akademischen Grades eines

DOKTOR-INGENIEURS

von der KIT-Fakultät für
Elektrotechnik und Informationstechnik
des Karlsruher Instituts für Technologie (KIT)

genehmigte

DISSERTATION

von

M.Sc. Simon J. Otten
geb. in Zwolle, die Niederlande

Tag der mündlichen Prüfung: 25. Januar 2019

Hauptreferent: Prof. Dr.-Ing. Mathias Noe
Korreferent: Prof. Dr. Ir. Herman ten Kate

Acknowledgement

I am grateful to my supervisors Anna Kario, Wilfried Goldacker and Mathias Noe, who have supported me at all times, and from whom I have learned a lot. In frame of the EuCARD-2 project, a number of joint tests were done with valuable results. For this I thank Peng Gao, Marc Dhallé, Sander Wessel and Bas van Nugteren from Twente University and Maria Durante and Clément Lorin from CEA Saclay. Finally, I would like to thank my colleagues at the Institute for Technical Physics for the scientific discussions and contributions to the experiments, which were indispensable to this work: Karoline Altmann, Nadezda Bagrets, Eduard Demenčik, Antje Drechsler, Reinhardt Ernst, Holger Fillinger, Rainer Gehring, Francesco Grilli, Alexandra Jung, Andrea Kling, Andrej Kudymow, Mayraluna Lao, Uwe Mirasch, Rainer Nast, Bernd Ringsdorf, Brigitte Runtsch, Sonja Schlachter, Uwe Walschburger, Johann Willms, Michael Wolf and Hong Wu.

Contents

Chapter 1

Introduction

This thesis describes a number of studies on high-temperature superconducting Roebel cables. These are part of an effort to develop Roebel cables for high-field magnets, which have potential for use in a new generation of particle accelerators. The first chapter contains the necessary background information about superconductivity and Roebel cables, as well as the motivation for the research in this dissertation.

1.1 Practical superconductors for use in high-field magnets

Superconductors are materials characterized by zero electrical resistivity and expulsion of the magnetic field. A superconductor can carry a limited current density, called the critical current density J_c. The critical current density depends on both the temperature T and the magnetic field B, and becomes zero if either the critical temperature T_c or the critical magnetic field B_c is reached. The boundaries of superconductivity can be visualized by a critical surface, as illustrated in figure 1.1. A superconductor is in the superconducting state only if the coordinate given by the current density, magnetic field and temperature lies below this surface.

In 1911, superconductivity was observed for the first time in mercury [1], which has a critical temperature of 4.2 K [2]. Over the years, superconducting properties were found in many other metals and compounds with increasing critical temperatures. Several of these materials are used in practical superconducting wires that are suitable for long-length production and the winding of coils. A chart with the current carrying capacity of state-of-the art practical superconductors is maintained by P. Lee [3]. This chart (figure 1.2) shows the critical current density of the whole wire at 4.2 K, the temperature of a boiling liquid helium bath at atmospheric pressure. These superconducting wires can carry currents densities exceeding 1000 A/mm^2 with negligible dissipation. This makes them suitable for the construction of high-field superconducting magnets, which are more compact and have lower power usage compared to copper electromagnets. Applications of superconducting magnets include magnetic resonance imaging (MRI), nuclear magnetic resonance (NMR) spectroscopy, particle accelerators and nuclear fusion reactors.

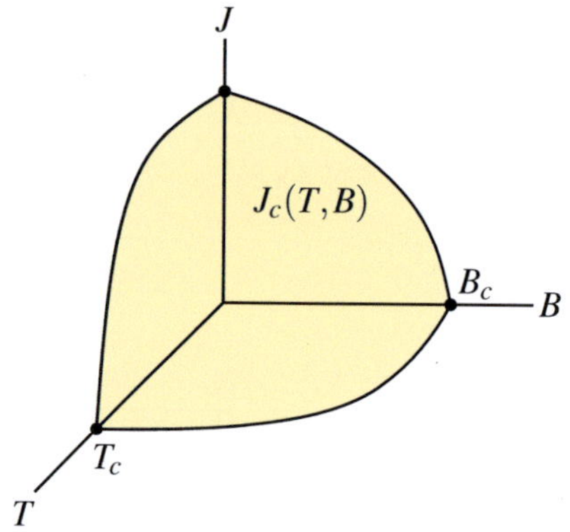

Figure 1.1: The critical surface of a superconducting material. The critical current density J_c is a function of the magnetic field B and temperature T. The superconductor can carry a current with zero resistivity as long as it does not exceed the critical value.

Most superconducting magnets are wound from wires containing niobium-titanium (NbTi) or niobium-tin (Nb_3Sn). These materials are called low-temperature superconductors (LTS), because their critical temperatures are low at 9.3 and 18.3 K respectively. Niobium-titanium is a ductile alloy which makes it easy to deform and it is relatively inexpensive. Its critical current density, however, drops strongly with increasing magnetic field, limiting the maximum field at 4.2 K to roughly 9 T. Niobium-titanium wires have been the conductor of choice for small research magnets as well as many larger projects. For instance, NbTi is used in the Large Hadron Collider (LHC) at CERN [4], where superconducting magnets focus the proton beam and guide it through the accelerator. Another example is the nuclear fusion reactor Wendelstein 7-X, in which a system of NbTi superconducting magnets is used to contain the plasma [5]. To attain magnetic fields exceeding 9 T, superconducting wires containing Nb_3Sn have been developed. Due to the brittleness of Nb_3Sn, such wires are more complex in fabrication. For this reason, Nb_3Sn wires are usually applied only if the magnetic fields exceed 9 T and NbTi wires cannot carry the required current. Wires with Nb_3Sn are extensively used in the experimental fusion reactor ITER, in which the central solenoid produces a field of 13 T [6, 7]. They are also used in the quadrupole magnets of the high-luminosity upgrade of the LHC, which focus the particle beam before colliding [8].

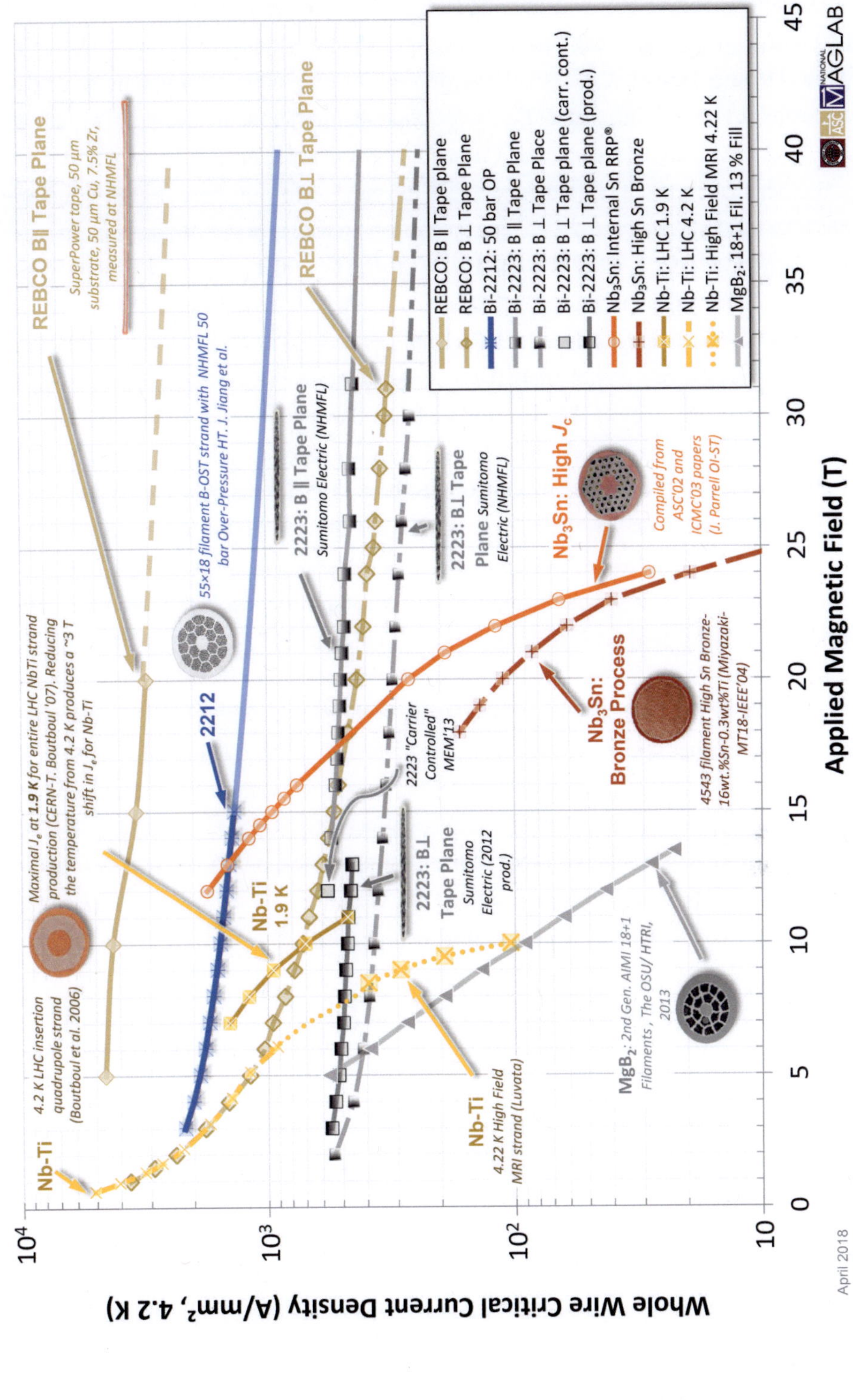

Figure 1.2: Whole wire critical current density of practical superconductors as a function of magnetic field at $T = 4.2$ K. Chart by P. Lee [3].

In 1986, superconductivity was observed in a BaLaCuO system at an unusually high temperature of 30 K [9]. One year later, a YBaCuO sample was found to be superconducting with a critical temperature of 92 K [10]. These discoveries were a breakthrough in the field of superconductivity, because the existent Bardeen-Cooper-Schrieffer theory could not explain superconductivity at such high temperatures. Unusually high critical temperatures were discovered in several other materials containing copper and oxygen. A few of these high-temperature superconductors (HTS) have been used in practical superconducting wires. Round and flat wires containing $Bi_2Sr_2CaCu_2O_{6+x}$ (Bi-2212) and $Bi_2Sr_2Ca_2Cu_3O_{8+x}$ (Bi-2223) have been prepared. The high-temperature superconductor *RE*BCO, which stands for rare-earth metal barium copper oxide, is applied in flat coated conductors (CC). Until now, no usable *RE*BCO round wires have been prepared, because they do not attain the required alignment of grains (see section 1.2.1). The current carrying capacities of HTS wires at 4.2 K are shown in figure 1.2. Compared to LTS wires, the critical current densities of HTS wires have a weaker dependence on the magnetic field. At fields above 20 T, they outperform even the best Nb_3Sn wires by a large margin.

HTS wires have led to the development of a new generation of high-field superconducting magnets. Using HTS wires, record fields for all-superconducting magnets of 26.4 T [11] and 27.0 T were reached [12]. High-temperature superconductors are also considered for use in large scale applications, such as fusion reactors [13, 14] and particle accelerators [15]. For large coils, single superconducting wires cannot carry enough current. A very large number of windings would be needed to reach the required field, leading to an excessive inductance. Such a magnet could be ramped only slowly or under a high voltage, which is undesirable for stability reasons. Instead, large magnets are constructed from superconducting cables, which contain multiple wires connected in parallel. In LTS accelerator magnets, flattened helical cables called Rutherford cables are commonly used [16]. Cable-in-conduit conductors carrying currents up to 82 kA are developed for use in fusion reactors [17]. Unfortunately these round-wire cabling techniques cannot be used for HTS wires because, except for Bi-2212 wires, they are flat tapes.

A number of new cabling techniques for flat *RE*BCO coated conductors is currently being developed. Of these, three concepts are the most promising for the use in high-field magnets: cables containing a twisted stack of coated conductors [18, 19], coated-conductor-on-round-core (CORC) or tube (CORT) cables [20, 21] and Roebel cables [22–24]. The latter has been developed at the Institute for Technical Physics (ITEP) for about 10 years, and is the subject of research in this thesis. Roebel cables are prepared by cutting coated conductors into a meandering shape, followed by assembly of multiple strands into a cable. A picture of a Roebel cable is shown in figure 1.3. Roebel cables have a high packing fraction, are fully transposed and retain the mechanical flexibility of single coated conductors. The preparation of *RE*BCO coated conductors and Roebel cables are be described in more detail in the next section.

Sections 1.3 and 1.4 discuss the potential applications of Roebel cables and the motivation for the investigations described in this thesis.

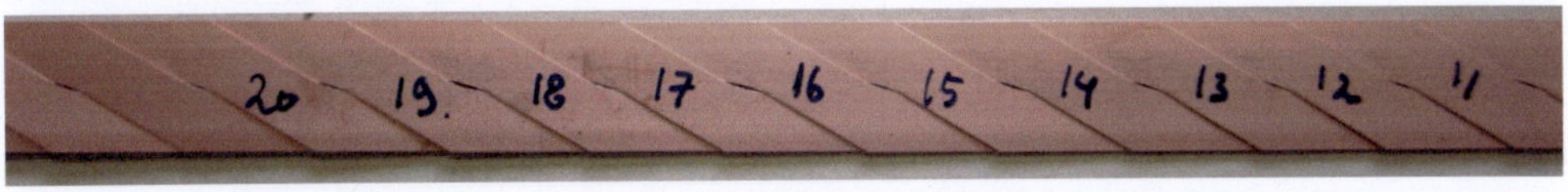

Figure 1.3: Picture of a Roebel cable assembled from patterned coated conductors.

1.2 Preparation of *RE*BCO Roebel cables

1.2.1 *RE*BCO coated conductors

In polycrystalline *RE*BCO, the macroscopic critical current density depends on the misorientation angle at grain boundaries. It decreases by a factor 50 if the misorientation angle is larger than 20° [25]. For this reason, polycrystalline *RE*BCO wires with randomly oriented grains have poor current carrying capacity. To improve the current carrying capacity, *RE*BCO coated conductors (CC) have been developed. By depositing *RE*BCO on a textured substrate, the misorientation angles are kept below 5°. Grain alignment is attained using a "rolling assisted biaxial textured substrate" (RABiTS), or by depositing a textured buffer on the substrate using ion-beam assisted deposition (IBAD) [26].

Figure 1.4 shows a cross-section of a typical coated conductor. The substrate is made of a stainless steel or nickel alloy[1] tape and has a thickness between 30 and 100 µm. A stack of buffer layers is deposited on the substrate. In IBAD coated conductors, a textured MgO buffer layer ensures grain alignment. A 1-5 µm thick layer of superconducting *RE*BCO is deposited on the buffer layers. This can be done using various techniques including metal-organic deposition (MOD) [27] or pulsed laser deposition (PLD) [28]. A 1-5 µm layer of silver is sputtered on top of the *RE*BCO layer for chemical protection. Additionally, the silver layer stabilizes the conductor by providing an alternative path for the current at weak spots in the superconductor. Depending on the application, a copper layer of up to 100 µm thickness is added for additional stabilization. The copper layer can either be soldered onto the tape or formed by electro-deposition.

[1] For example Monel or Hastelloy C-276

*RE*BCO coated conductors have been produced in lengths exceeding 1000 m [29]. Companies currently preparing coated conductors include American Superconductor (AMSC) [30], Bruker HTS [31], Deutsche Nanoschicht [32], Fujikura [33], Oxolutia [34], Shanghai Superconductor Technology [35], Shanghai Creative Superconductor Technologies [36], SuNAM [37], Superconductor Technologies inc. (STI) [38], SuperOx [39], SuperPower [40] and Theva [41].

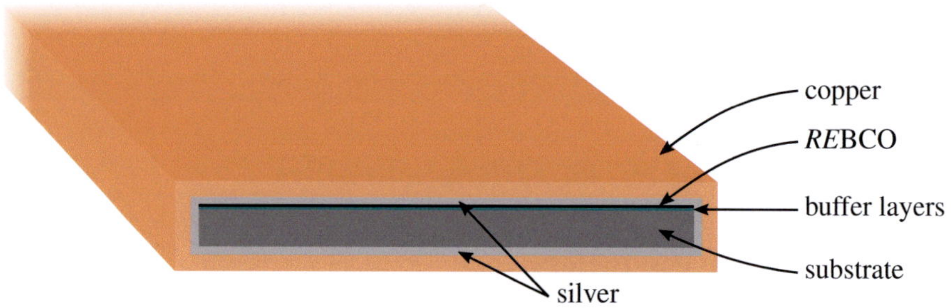

Figure 1.4: Schematic cross-section of a *RE*BCO coated conductor

1.2.2 Roebel cables

The Roebel bar was invented by Ludwig Roebel in 1912 as a means to reduce eddy currents in large generators.[2] A Roebel bar consists of insulated copper strands transposed in a helical way. A similar concept is used in superconducting Roebel assembled coated conductors (RACC), which for brevity are called Roebel cables in this thesis. The strands of a Roebel cable are prepared by cutting *RE*BCO coated conductors in a meandering shape. The cutting pattern is shown in figure 1.5. The geometry of the pattern is defined by parameters for the width of the straight section w_s, the width of the cross-over w_c, the transposition length ℓ_t, the cross-over angle α and the inner radius R_{in}. The pattern is repeated each transposition length. At ITEP, the coated conductors are cut using a pneumatic punching tool. This tool and considerations about punching quality are discussed in appendix A.

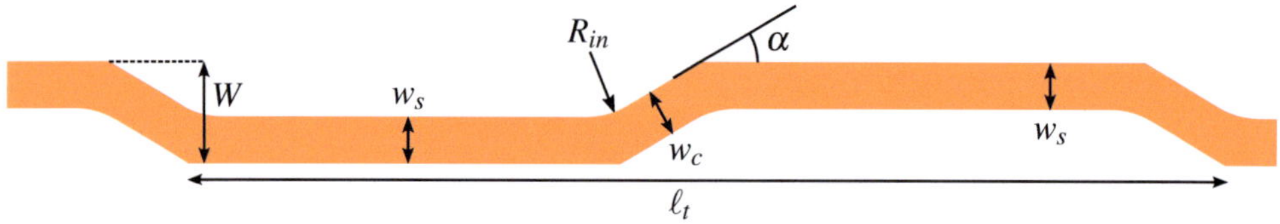

Figure 1.5: The Roebel cutting pattern with defining parameters. W, w_s, and w_c are the widths of the entire tape, the straight part and the cross-over. α is the cross-over angle, R_{in} is the inner radius and ℓ_t is the transposition length.

[2] Der Roebelstab von Ludwig Roebel (1878 - 1934):
`https://www.mannheim.de/de/wirtschaft-entwickeln/roebelstab-ludwig-roebel-1878-1934`

Industrial coated conductors usually have a width of 4 or 12 mm. Punching tools suited to both widths are available at ITEP. The geometry parameters for both tools are listed in table 1.1. The tool for 12-mm-wide tapes consists of two movable sets of punches and dies, which can be fixed in different positions. This enables punching of strands with different widths and transposition lengths. The tool for narrow tapes (4 mm) has a fixed geometry.

Multiple punched strands are assembled to make a Roebel cable. The strands are added one-by-one, by winding them around the cable in a helical way. For short cables, this is done by hand. For longer cables, a semi-automatic cable assembly machine is used. The assembly process is illustrated in figure 1.6. The strands in the assembled cable are fully transposed, which means that each strand moves through all positions within the cable in the same way. The full transposition ensures that each strand has the same path length if the cable is bent. This makes the Roebel cable easier to bend than an untransposed stack. The full transportation also makes sure that each strand has the same self-inductance and mutual-inductance with other strands. When the current is ramped, this will result in an equal current distribution between strands. This is desirable for applications where the field quality is important, such as accelerator magnets.

1:
2:
3:
4:
5:
6:

Figure 1.6: Assembly of a Roebel cable with six strands.

The maximum number of strands that can be added to the cable is determined by the transposition length. In case of a short transposition length of 126 mm, only 10 strands can be assembled. If a longer transposition length of 426 mm is used, up to 31 strands can be assembled.

Table 1.1: The possible geometry parameters using currently available punching tools. Tools are available both for 4 mm wide tapes (narrow) and 12 mm wide tapes (wide). The narrow tool has a fixed geometry while that of the wide tool can be changed.

parameter	narrow tool (fixed)	wide tool (changeable)	unit	description
W	4	12	mm	tape width
w_s	1.9	5.5/5.9	mm	width of the straight part
w_c	1.9	5.5/5.9	mm	width of the cross-over
α	30°	30°		cross-over angle
ℓ_t	116	126/226/300/426	mm	transposition length
R_{in}	1	10	mm	inner radius

1.3 Applications of Roebel cables

The Roebel cable has been selected for a number of prototype devices for various reasons. Industrial Research Limited constructed a superconducting transformer, in which a Roebel cable was chosen for the low-voltage winding because of its high current capacity and limited AC loss [42–44]. Roebel cables have been considered for use in aircraft power systems, where its low weight is beneficial [45]. Roebel cables can also be used as the strands of a larger transposed cable for very large devices such as nuclear fusion magnets [46].

In the European project EuCARD-2[3], several HTS accelerator-type dipole magnets are built [47]. For two designs, the aligned block coil (CERN) [48] and the cos-theta coil (CEA Saclay) [49], the Roebel cable has been selected as a conductor because of its high current density and full transposition. In the aligned block coil (figure 1.7a), the cables are aligned in such a way that the magnetic field is parallel to the cable surface. In this orientation, the critical current depends less strongly on the magnetic field, and thus a higher current density can be reached. CEA's cos-theta magnet (figure 1.7b) uses a more conventional layout that was also used for many low-temperature superconducting dipole magnets. Both coils are designed to generate a magnetic field of 5 T when operated stand-alone in a liquid helium bath ($T = 4.2$ K). As a next step, the magnets are to be tested in a 13 T background field of a larger LTS dipole magnet. In these conditions, the cos-theta and aligned block coils are designed to lift the magnet field to 15.5 T and 16.9 T respectively. These fields would be a record for any superconducting dipole magnet. The role of ITEP in this project is to deliver the Roebel cables with lengths up to 32 m, longer than any cables prepared here before. The development of these magnets has raised a

[3] European Coordination for Accelerator Research and Development 2 (`http://eucard2.web.cern.ch`)

number of questions about the mechanical, electromagnetic and thermal properties of Roebel cables, which are a motivation for the investigations in this thesis.

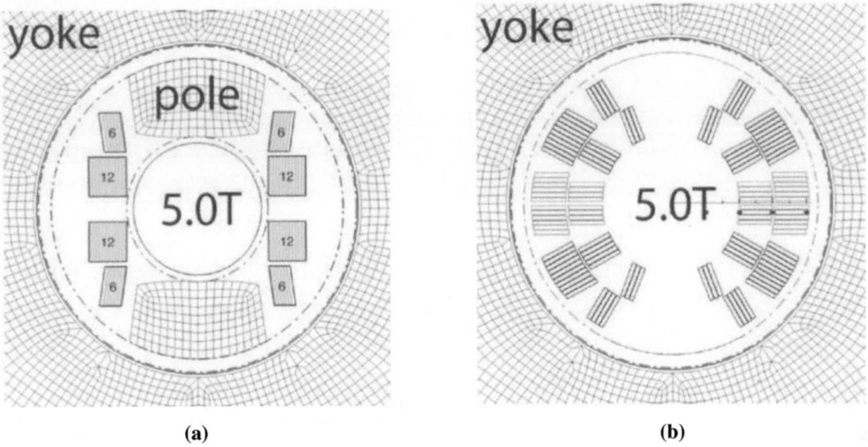

Figure 1.7: Cross-sections of the EuCARD-2 aligned-block coil (a) and cos-theta coil (b). Images by Kirby et al. [48].

1.4 Motivation and thesis contents

This thesis consists of two parts. In the first part, mechanical properties of *RE*BCO coated conductors and Roebel cables are investigated. The mechanical properties need to be understood for coil applications, because the winding process requires bending and torsion of the conductor. Additionally, the cables are subjected to tension and compression resulting from Lorentz forces. A number of experiments was done in order to simulate the mechanical stresses on short cable samples. These experiments are discussed in chapter 2.

The second part of the thesis deals with the effect of inter-strand resistance on the cable properties. Roebel cables with low resistance between strands allow current redistribution. If a local increase in temperature leads to a loss of superconductivity in one of the strands, the current can redistribute to other strands reducing further Joule heating (see figure 1.8). This mechanism has been shown to improve the stability of LTS Rutherford cables [50, 51]. However, it has not been investigated for HTS Roebel cables yet.

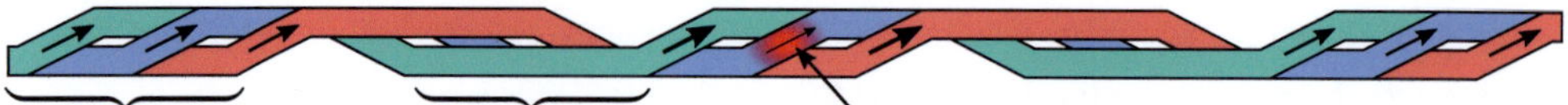

Figure 1.8: Current redistribution can decrease the current through a local increase of temperature. For clarity, only three strands are shown.

A low inter-strand resistance enables coupling currents to be induced by a time-dependent magnetic field. The path of coupling current induced by perpendicular magnetic field are sketched in figure 1.9. These coupling currents flow through the resistive interfaces between strands and cause dissipation. This effect is known as AC coupling loss. Coupling loss is undesirable as additional cooling power is needed and because the resulting temperature increase can lead to stability issues. The effect of inter-strand resistance on both stability and AC loss needs to be understood before the inter-strand resistance can be optimized.

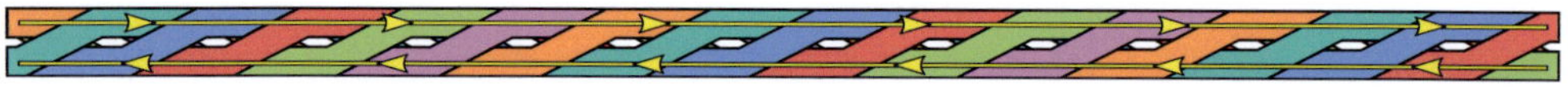

Figure 1.9: Coupling currents induced by a time-dependent magnetic field perpendicular to the wide face of the cable.

In chapter 3, a method to describe the inter-strand resistance using two parameters is proposed. The inter-strand resistance is measured in a number of cable samples under transverse pressure. AC loss measurements of cables with different levels of inter-strand resistance are described in chapter 4. The measured AC losses are compared to calculations. In chapter 5, the effect of inter-strand resistance on the stability is investigated. This is done by calculating the minimum quench energy (MQE) taking into account current redistribution between strands. The calculations are compared to an experimental MQE measurement on a short sample in quasi-adiabatic conditions.

Chapter 2

Mechanical characterization

2.1 Mechanical loads in the EuCARD-2 demonstrator dipole magnets

The demonstrator dipole magnets of the EuCARD-2 project are among the most advanced applications of Roebel cables. CERN develops an aligned block coil, in which the surface of the cable is aligned with magnetic field, while CEA Saclay uses a more conventional cos-theta layout (see section 1.3). Because the cables carry a large current density in a high magnetic field, they endure large mechanical stresses resulting from Lorentz forces. The calculated mechanical loads [52, 53] are listed in table 2.1.

The compressive stresses in the cos-theta coil resulting from pre-stress, cool-down and Lorentz forces were calculated by CEA using a finite element code [49]. Initially, a cos-theta coil with two concentric layers of Roebel cables was foreseen [49]. However, the calculated mechanical stresses were found to be excessively high with peak values up to 800 MPa. Such stresses exceed the maximum tolerable stress for a *RE*BCO coated conductor [54] and the structural materials in the coil. An adapted cos-theta design was developed with only one layer and a larger support structure [52], of which the expected compressive stress values are shown in table 2.1. The largest stress component is in the azimuthal direction, which in this design is perpendicular to the wide face of the cable. This stress is expected to peak at 55 MPa in stand-alone operation and 220 MPa in a 13.0 T background field.

In the aligned block coil, the wide face of the cable is aligned parallel to the magnetic field. Therefore, the Lorentz forces will be largely perpendicular to the wide face. The calculated peak value of this component is 110 MPa [53], and the stresses are therefore somewhat lower than in the cos-theta coil. These values were computed assuming that the cable is a strip with homogeneous mechanical properties. Stress concentrations that occur due to the internal structure of the cable are therefore not taken into account.

Table 2.1: Bending radius and computed compressive stress for the cos-theta design [52] and the aligned block design [53]. The compressive stress is separated in components perpendicular and parallel to the wide face of the cable. Stress levels are shown for stand-alone operation and operation in a 13 T background field.

	cos-theta design		aligned block design	
Minimum bending radius [mm]				
out-of-plane	7.5		16	
in-plane	200		200	
Operation mode	stand-alone	13 T background field	stand-alone	13 T background field
Background field [T]	–	13.0	–	13.0
Dipole field [T]	5.0	15.5	5.0	16.9
Current [kA]	11.7	7.7	7.9	8.1
Peak compressive stress [MPa]				
perpendicular	55	220	17	110
parallel	15	63	–	–

This chapter describes mechanical experiments that were done on Roebel cables. The aims were to develop a better understanding of the mechanical properties of Roebel cables, and to evaluate their suitability for the EuCARD-2 magnet designs. Section 2.2 discusses transverse loading experiments that were done to find the highest tolerable transverse stress and to determine whether the calculated values for the EuCARD-2 coils are acceptable. This work was done as a collaboration between KIT and the University of Twente. In section 2.3, out-of-plane bending experiments of different coated conductors and several Roebel cables are described. Section 2.4 describes a test that was done with CEA Saclay to qualify the coil exit of the inner winding, where the cable sees a twist and two S-shaped bends.

2.2 Transverse stress tolerance of Roebel cables with and without impregnation

This section is based on an article published in Superconductor Science and Technology [55].

2.2.1 Introduction

Transverse pressure tests on single *RE*BCO tapes show that they typically tolerate transverse stresses of at least 300 MPa [54, 56–58], exceeding the expected stresses in accelerator magnets. In a Roebel cable, however, the force is not homogeneously distributed, leading to local stress concentrations. J. Fleiter et al. showed that the effective stress bearing section of a Roebel cable made at ITEP on a flat anvil is only 23% [59]. This means that local peak stresses are at least four times higher than the average pressure. D. Uglietti et al. pressed Roebel cables with insulated strands to transverse stresses up to 70 MPa [57]. Degradation was observed at pressure levels as low as 10 MPa and the I_c value of most strands degraded by more than 20% at 40 MPa. These results indicate large stress concentrations and confirm the need to reinforce the cable.

A common reinforcement method used with resistive as well as superconducting coils is impregnation with epoxy resin. For *RE*BCO coated conductors, however, epoxy impregnation leads to complications. An impregnated *RE*BCO pancake coil from T. Takematsu et al. showed degradation of the critical current [60]. After visual inspection, a separation of the layers within the individual conductor (delamination) was observed. This type of damage results from a mismatch in thermal expansion between the conductor and the epoxy: When cooled down from room temperature to liquid helium temperature ($T = 4.2$ K), epoxy seeks to contract by 1.33%, while the *RE*BCO tape contracts by only 0.25% [61]. This mismatch leads to thermal stresses which can delaminate the conductor.

Several different methods to reduce such thermal stresses on the conductor have been proposed and were tested successfully. Epoxy resin bonds firmly to metals. Impregnation materials with a lower bonding strength cannot build up as much tensile stress on the surface. T. Takematsu et al. impregnated a *RE*BCO pancake coil with paraffin, which has a negligibly small bonding strength [60]. The critical current of this coil was not affected by the impregnation. M. Matsumoto et al. achieved a field of 24 T with *RE*BCO pancake coils impregnated using this method [62]. However, impregnation with paraffin might not be suitable for Roebel cables because it may lack the mechanical strength needed to reduce stress concentrations. K. Mizuno et al. proposed impregnation of *RE*BCO pancake coils with cyanoacrylate resin [63]. They showed that this resin has a bonding strength significantly lower than epoxy, and demonstrated that a coil can be impregnated with this material without degradation of the critical current. Tensile stresses can also be reduced by introducing a weak mechanical barrier between the conductor and the epoxy resin. T. Trociewitz et al. did this by inserting the *RE*BCO coated conductor in a polyester heat-shrink tube before coil winding and impregnation [64]. A layer-wound coil produced with this method generated a record magnetic field of 35.4 T. A similar concept was used by Y. Yanagisawa et al., who added a layer of polyimide to the *RE*BCO conductor by electro-deposition [65]. Also this method was successful in preventing degradation due to impregnation. A different approach is to add filler particles with low thermal expansion to the epoxy resin. C. Barth et al. impregnated a Roebel cable with a 1:1 mixture of epoxy and fused silica without critical current degradation at $T = 77$ K and in self-field [61].

In this work, we report on the transverse strength of short Roebel cables impregnated with such an epoxy-silica mixture. This impregnation method was chosen because of the proven mechanical properties of epoxy and its easy application to the cable structure using vacuum impregnation. Adding polyester or polyimide barriers to the strands or using resins other than epoxy may remain interesting options for future investigation. For comparison, a cable without impregnation was tested as well.

2.2.2 Experimental details

This section describes the preparation of Roebel cables for the transverse cable press at the University of Twente. In this facility, forces up to 250 kN can be applied using an electromechanical press. The sample current is generated by a 50 kA superconducting transformer. Both transformer and press were built earlier and used in transverse stress tests of LTS Rutherford cables. A schematic representation of the facility is shown in figure 2.1; more details can be found in [66].

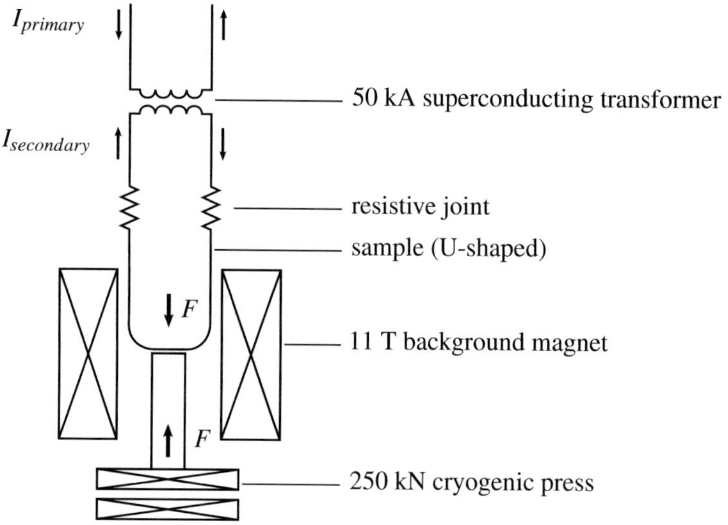

Figure 2.1: Scheme of the test facility for superconducting cables at the University of Twente [67].

Three identical 75 cm long 10-strand cables were prepared from *RE*BCO tape from SuperPower (SCS12050-AP). A 12 mm wide tape with a 50 μm Hastelloy substrate, 2 μm of silver and 40 μm of copper stabilization was used. The tape was punched into the meandering Roebel pattern with a transposition length of 126 mm [22]. In this process, the current carrying width of the tape is reduced from 12 to 5.5 mm. Before cable assembly, the critical current of the separate strands was measured in liquid nitrogen ($T = 77$ K) and magnetic self-field. The strands had an average I_c value of 172 ± 2 A. This corresponds to a critical surface current density of 31 A/mm-width compared to 33 A/mm-width before punching. The average n-value[1] was 29 ± 1.

After assembly, the cable was mounted on a U-shaped sample holder, suitable for measurements inside a solenoid magnet (figure 2.2). The corners of the sample holder have a radius of 20 mm; the horizontal section in between the bends is 26 mm long. The *RE*BCO coated sides of all strands are facing the sample holder, since this side will be soldered to the current leads. The cable is electrically insulated from the stainless steel sample holder with a layer of polyimide tape. A block of Teflon is pushed against the flat "bottom" of the U-shape to create a flat epoxy surface and to prevent epoxy from flowing out during the curing process. Three pairs of voltage taps are soldered to the cable over one transposition length including the bottom sample section.

Two of three cables (cable 1 and 2) were vacuum impregnated with a mixture of epoxy and fused silica powder. The epoxy resin Araldite CY5538 with hardener HY5571 was supplied by Huntsman corporation, and it is mixed with Silbond FW600 EST fused silica powder with a median particle size of 4 μm. The mixing ratio of resin, hardener and silica powder is 1:1:2 by weight. To remove trapped gas bubbles, the mixture is evacuated to 1-2 mbar for 30 minutes. Impregnation is done by slowly lowering the sample holder into an epoxy bath at low pressure (3 mbar), and then releasing the vacuum. During impregnation, the epoxy bath as well as the sample holder are heated to 80 °C to reduce viscosity. The sample is retracted from the bath and the resin is cured at 100 °C for 24 hours.

After curing the Teflon block is removed. An impregnated cable in this state is shown in figure 2.3a. Next, two layers of glass fibre ribbon wetted with Stycast 2850FT/23LV epoxy are added to the cable. The 30 mm long pressure anvil is positioned on top of the glass ribbons. The surface of the anvil is aligned at a distance of 1.5 mm from the sample holder using two positioning plates (figure 2.3b), which are removed when the Stycast has hardened. The cables with the block glued in place are shown in figure 2.3c. The anvil is glued for two reasons: To avoid stress concentrations at the ends of the pressed section and to ensure parallelism of the anvil and the sample surface. The layer of Stycast and glass fibre between the cable and the

[1] The n-value is a measure of abruptness of the transition between superconducting and resistive states. See section 5.2 for more information.

stainless steel anvil is relatively soft, while the anvil is aligned with the sample holder. Both precautions ensure that the force is transferred homogeneously over the entire surface covered by the anvil.

A drawback of this preparation method is the possibility of a bond between the anvil and the plates that prevent sample motion under influence of the lateral Lorentz forces (figure 2.3a). The support plates are fixed to the sample holder and may therefore transmit part of the transverse force during the actual experiment. To minimize this complication, the sides of the pressure anvil and inner surface of the support plates were covered with Kapton tape which hardly bonds with the Stycast resin and sticks to the metal plates with a relatively weak silicone adhesive. As a verification that the lateral support plates play a negligible role in the pressure tests, the support plates were made lower for cable 2 (3 mm vs. 14 mm above the sample holder), in order to reduce the contact area with the anvil.

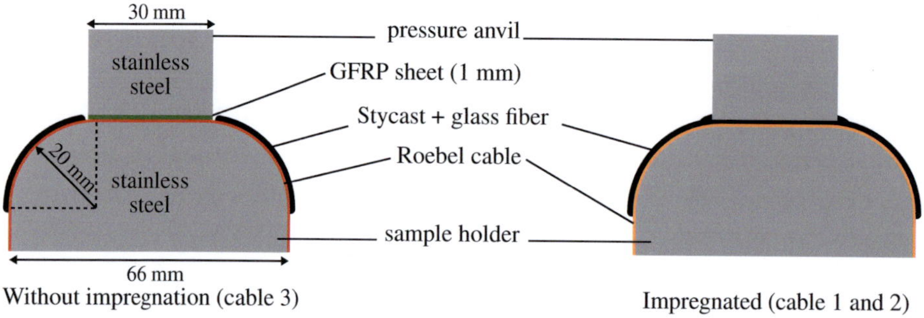

Figure 2.2: Cross-sectional view of the experiment for both samples with and without impregnation.

(a) (b) (c)

Figure 2.3: Cable 2 on the U-shaped sample holder. The cable is supported against the lateral Lorentz forces on both sides by side plates (1). (a) After impregnation and removal of the Teflon block. (b) The pressure anvil (2) being glued in place with Stycast epoxy. The block is aligned to the sample holder using two positioning plates (3). (c) After curing the Stycast and removal of the positioning plates.

Unlike cables 1 and 2, sample 3 was not impregnated, and the pushing block was not glued to it. Instead, a 1 mm thick sheet of glass-fibre reinforced plastic (GFRP) was attached to the side of the anvil in contact with the cable. This sheet compensates for the thinner sample, and reduces stress concentrations at the ends of the pressed section, where otherwise the corners of the anvil would directly cut into the cable. This results in a fairer comparison with impregnated cables, where the anvil and the sample were separated by a 1 mm thick layer of glass fibre with Stycast. Since the cable is not impregnated, it is necessary to reinforce it against Lorentz forces in another way. Outside of the pressed section, along the corners of the U-shape, four layers of glass fibre soaked in Stycast 2850FT epoxy were added. The uncovered section of the cable is supported against Lorentz forces by applying a pressure of 10 MPa before doing any measurements.

Finally, the sample is connected to the NbTi current leads of the transformer by soldering over one transposition length (126 mm) with $Sn_{97}Ag_3$ solder. Wires for the voltage measurement are connected to three strands with a separation of 126 mm. All these three strands have a cross-over in the pressed segment. Voltage measurements on the seven remaining strands are not possible, because these strands cannot be reached from the cable surface between the current contact and the pressed segment. A fourth pair of voltage taps is connected to the current contacts.

2.2.3 Results and discussion

All current-voltage (I-V) measurements were done at $T = 4.2$ K in a perpendicular applied magnetic field $B_{\perp} = 10.5$ T. The initial I-V curves for the three cables are shown in figure 2.4. In cables 1 and 3, voltage measurements on several strands did not yield useful data and are therefore not shown in the figure.

The critical current of the used coated conductors strongly depends on the orientation of the magnetic field, and is lower if the magnetic field is perpendicular to the cable surface. A voltage will therefore first appear in the cable section where the angle between the wide cable surface and the magnetic field is close to 90°. As the straight section of the cable is relatively short compared to the bends, this length is less well-defined. For determining the critical current, a straight section length of 30 mm is used, corresponding to the length of the anvil. All possible damage due to transverse pressure will occur in this segment. Using this definition and an electric field criterion of 10^{-4} V/m, the voltage criterion becomes 3 μV.

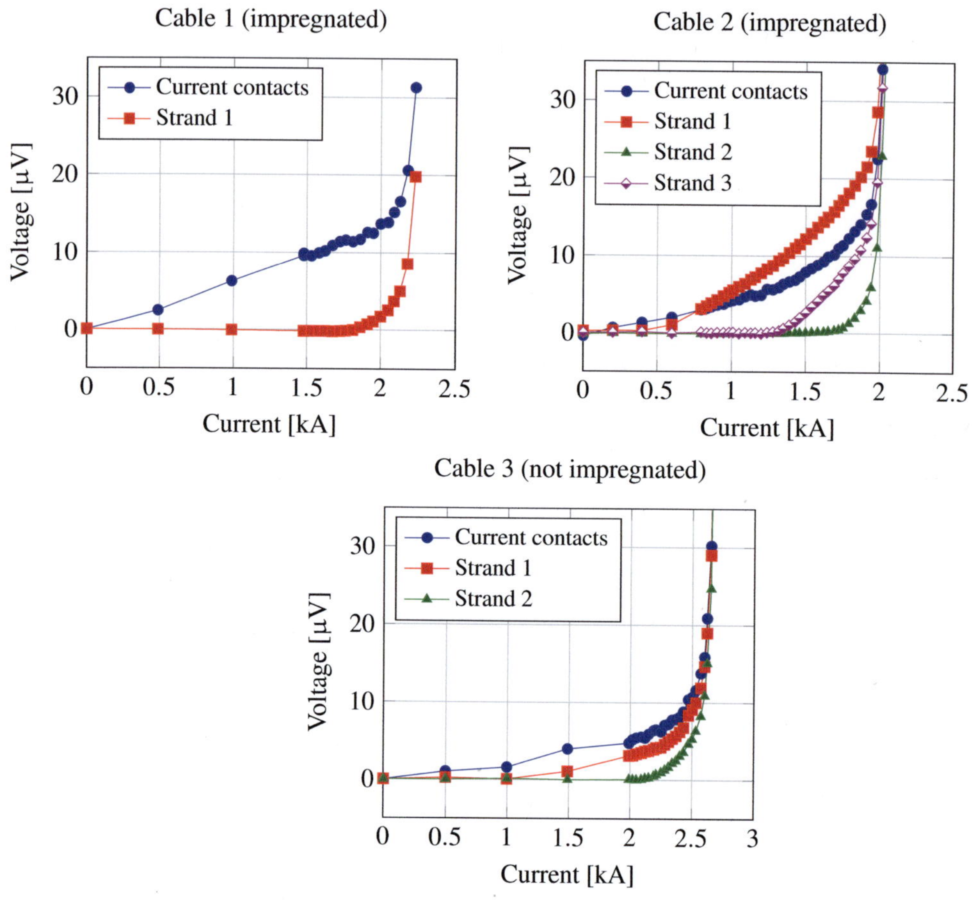

Figure 2.4: Initial IV curves of all cables at $T = 4.2$ K, $B_{\perp} = 10.5$ T. The voltage was measured over the current contacts and over strands including the pressed section.

As discussed in [68], current (re)distribution effects in short-sample measurements lead to the appearance of premature voltages. Determining the critical current by interpolation at 3 µV would result in scattered values for the different voltage signals, which would not be representative for the critical current of the entire cable. For this reason, the critical current is computed by fitting a voltage-current power law only through measurements points obtained at higher currents, at which all signals converge. This occurs at current corresponding to a voltage of approximately 20 µV. This method yields consistent critical currents (within 60 A) from the different voltage signals.

In the impregnated cables, the initial critical currents were 2.07 kA for cable 1 and 1.87 kA for cable 2. Cable 3, which was not impregnated, had a markedly higher critical current of 2.53 kA. This may indicate that despite the fused silica filler, epoxy impregnation still causes some damage.

However, the impregnated cables are clearly less sensitive to transverse stress. Figure 2.5 shows the critical currents as a function of the transverse stress. The stress is calculated dividing the measured force by the 12 mm × 30 mm area of the anvil. The critical current of cable 3 started to decrease at stress levels as low as 40 MPa. The impregnated cables on the other hand were not affected for stress levels up to 254 MPa for cable 1 and 169 MPa for cable 2. There is a rather large difference in onset of degradation between the two impregnated samples. Some deviation between different samples is to be expected, since the exact geometry of impregnated cables is hard to control. At increasing stress levels though, there is no clear difference between the strength of the two samples. This demonstrates that a possible bond to the lateral support plates did not play a large role.

To determine whether the I_c degradation is reversible, several measurements were done after reducing the stress level. The critical current did not recover, so that the degradation is indeed irreversible. The total critical current degradation observed during the experiment was 27% at 327 MPa for cable 1, 4.5% at 357 MPa for cable 2 and 60% at 198 MPa for cable 3.

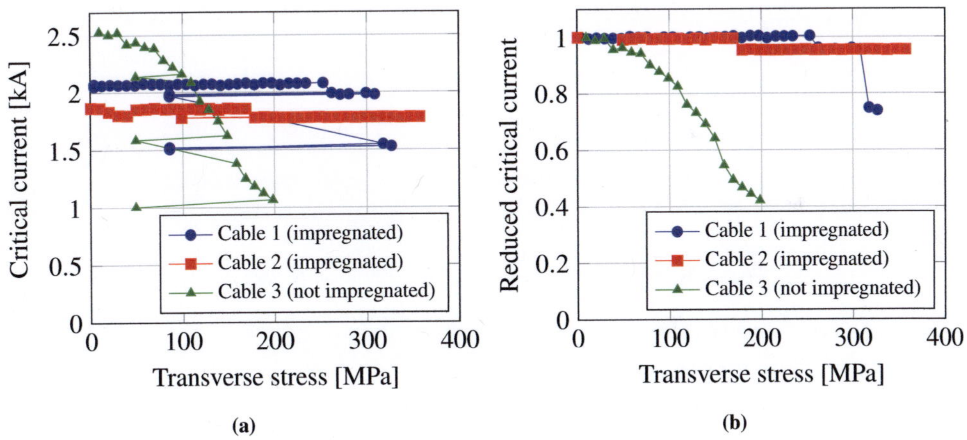

Figure 2.5: (a) The critical current as a function of transverse stress. The lines connect the data points in chronological order. (b) Critical current normalized to the initial value for each sample.

The measurements show that the impregnated cables withstand transverse pressures up to at least 169 MPa. This is a first confirmation that impregnated Roebel cables can withstand stress levels similar to those expected in accelerator magnets. A remaining point of discussion is that the pressed section of 30 mm is shorter than the transposition length of 126 mm. In the pressed section, only four out of the ten strands have a cross-over from one side of the cable to the other, which is the location where stress concentrations are expected to occur [57, 59]. As a result, the measured degradation might be lower than in the case of a longer pressure anvil. Arguably, however, this will not affect the point of onset of the degradation, as it does not influence the magnitude of the stress concentrations at each cross-over.

A cross-section was prepared from cable 1 to check the impregnation quality and the alignment of the anvil with the sample holder, and to inspect the cable for visible damage. The cable and the anvil to which it was glued were cast in epoxy. Material was then removed by sanding until the pressed section was visible. The surface was polished and examined under an optical microscope. Figure 2.6 shows the entire cross-section. The silica-epoxy mixture filled the structure throughout and there are no bubbles visible. This indicates a good impregnation quality. In the Stycast layer there are some empty spaces because no vacuum was used. The thickness of the whole cable-Stycast structure is close to 1.45 mm over the entire width. Including the 50 μm Kapton insulation on the sample holder, the distance between the sample holder and pressure anvil was 1.50 mm, as designed. Note that the impregnated cable by itself is thicker on the left than on the right, presumably because the Teflon block was not exactly straight during impregnation. The difference in height is corrected for by glueing the pressure anvil using the positioning plates. The only visible damage is delamination of the tape closest to the sample holder, clearly visible in figure 2.7. However, it is unclear whether this delamination is a result of thermal stresses or it occurred when the cable was removed from the sample holder, since for this some tensile force was needed.

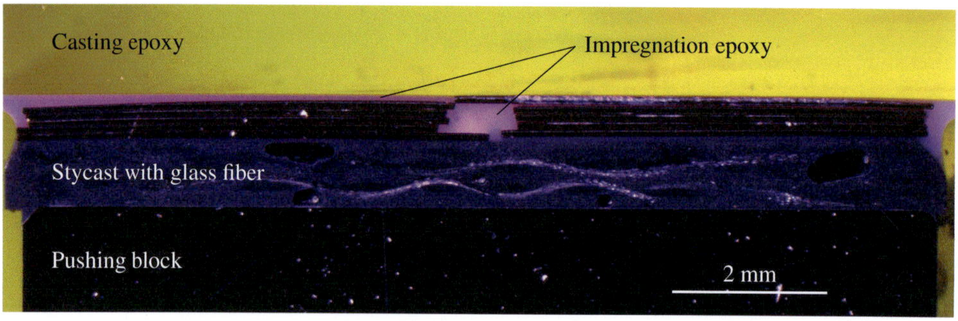

Figure 2.6: Cross-section of sample 1. The upper surface was placed on the sample holder. From top to bottom one can distinguish the casting epoxy; the Roebel cable; the glass-reinforced Stycast layer; and the pressure anvil.

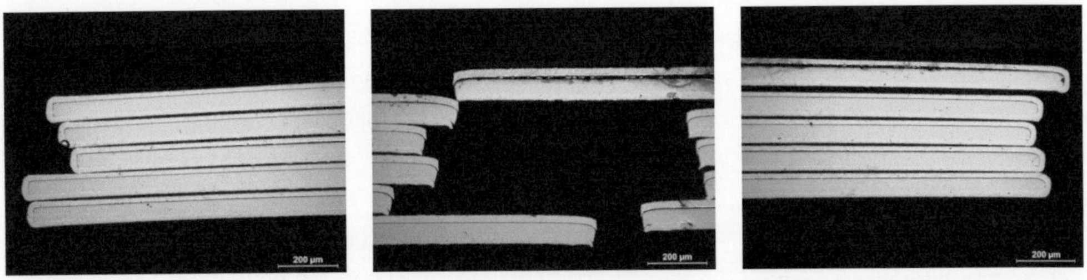

Figure 2.7: Microscopic images of the left end, the central hole and the right end of the cross-section. Delamination is visible in the upper right tape. The *RE*BCO layers are on the top side of each tape.

A marked increase was observed in the transverse stress tolerance of the Roebel cables after impregnation, constituting an important first step in demonstrating their use in accelerator magnets. In these experiments pressure was applied over relatively short lengths of just two impregnated samples, resulting in somewhat different stress-critical current characteristics. Also possible effects of cyclic loading, closely mimicking the mechanical conditions in repeatedly ramped magnet windings, remain to be addressed.

2.2.4 Conclusion

Two *RE*BCO Roebel cables (cable 1 and 2) were vacuum impregnated with a mixture of epoxy resin and fused silica powder. The critical current of the impregnated cables as well as a reference cable that was not impregnated (cable 3) was measured as function of transverse pressure at 4.2 K and in 10.5 T magnetic field. The initial critical current was 2.07 kA for cable 1, 1.87 kA for cable 2 and 2.54 kA for cable 3. The critical current reduction of the impregnated samples compared to the cable that was not impregnated may indicate damage due to impregnation. Pressure levels up to 357 MPa were applied over a length of 30 mm. No degradation was observed for pressures up to 253 MPa for cable 1 and up to 169 MPa for cable 2. In contrast, the critical current of cable 3 started to decrease already at stress levels as low as 40 MPa. These results are a first confirmation that impregnated Roebel cables can withstand transverse pressures well above the 110 MPa expected in accelerator magnets.

The work is currently being continued with more cable samples at the University of Twente. New tests are done to compare the impregnation method described above to a different impregnation method proposed by CERN. This proposed method uses an unfilled resin instead of a silica-filled one, and a glass-fibre sleeve is used to reduce the thermal expansion of the resin. Secondly, Roebel cables are tested made using material from Bruker, the partner in EuCARD-2 delivering conductors. P. Gao reported on the progress of this work [69].

2.3 Out-of-plane bending properties

This section is based on an article published in Superconductor Science and Technology [70].

2.3.1 Introduction

Construction of superconducting magnets requires deformation of the conductor. Deformations include bending, both in-plane and out-of-plane in case of a flat conductor, and torsion. The minimum out-of-plane bending radii in the EuCARD-2 demonstrator coils are relatively small: 16 mm in the aligned-block coil and 7.5 mm in the cos-theta coil. Such deformation induces strain in the conductor which can influence the current carrying properties. As in other superconductors, the reversible strain effect occurs in *RE*BCO [71]. Secondly, excessive strain can cause cracks in the superconductors leading to irreversible degradation of the critical current. Understanding of these effects is needed for coil design and testing.

The effect of bending on the current-carrying properties of coated conductors has already been investigated in a number of studies. In an early investigation by Usoskin et al. and Goldacker et al., minimum bending radii ranging from 8.5 to 15 mm were found [72, 73]. The bending behaviour was found to depend on whether the superconducting layer is on the inside or on the outside of the bend [72]. A reversible increase of the critical current under compressive bending was first observed by Sugano et al. [74]. In extensive work by Shin et al., bending properties were investigated in different bending modes [75], at different temperatures [76], and in tapes with different stabilizing layers [77, 78]. Recently, Shin et al. found that the irreversible bending limit is lower if the tape is subjected to cycled tension-compression bending [79].

More data on strain effects in coated conductors is available from investigations that induce strain using a method other than bending. These methods include the axial pull [77, 80], four-point bending [81–85], the U-spring [86, 87] and the Walters spring [88–90]. The reversible strain effect is found to depend on several parameters: van der Laan et al. found that the reversible strain is anisotropic in the ab-plane [85], and that the strain sensitivity of YBaCuO is higher than that of GdBaCuO [84]. In a work by Sugano et al., the effect of magnetic field and temperature on the reversible strain effect is investigated [80]. The strain sensitivity was found to increase with increasing temperature.

These findings show that the strain effect depends on many parameters. Bending properties can therefore not be assumed to be the same for tapes from different manufacturers. This is a motivation to include a larger number of different coated conductors in the study.

In this work, the effect of out-of-plane bending on coated conductors and Roebel cables is investigated at 77 K and in magnetic self-field. We provide a comparison of conductors from different manufacturers using the same measurement method. By choosing an appropriate bending sequence, reversible and irreversible effects are separated. The reversible effect at 77 K is of limited relevance for final coil testing at 4.2 K, since it depends on temperature and magnetic field. Irreversible effects are presumed to be related to actual damage in the superconductor. Such damage is always undesirable regardless of application. Therefore, those results are more relevant to the EuCARD-2 demonstrator coils.

2.3.2 Bending of *RE*BCO tapes

The single tapes are bent using a bending device developed by Goldacker et al. [91, 92]. The device can continuously bend the sample at low temperature ($T = 77$ K), enabling a high resolution in bending radius without the need for thermal cycling. A restriction of this method is that it is not fully representative for bending during coil winding, because this is done at room temperature under different pre-strain conditions.

A picture of the device is shown in figure 2.8. The sample is fixed on both ends in rotatable parts, which also function as current leads. By turning a rod, the angle between the two parts can be continuously changed from 0 to 180°, inside a liquid nitrogen bath. At 180°, the minimum bending radius of 5 mm is reached. The parts move in such a way that the sample shape is always close to a circle. This ensures homogeneous bending within a displacement error of 2% [18]. The length of the bent section is constant at $\ell = \pi * 5$ mm ≈ 15.7 mm. The bending radius at intermediate angles is given by $R = \ell/\alpha$, in which α is the bending angle in radians.

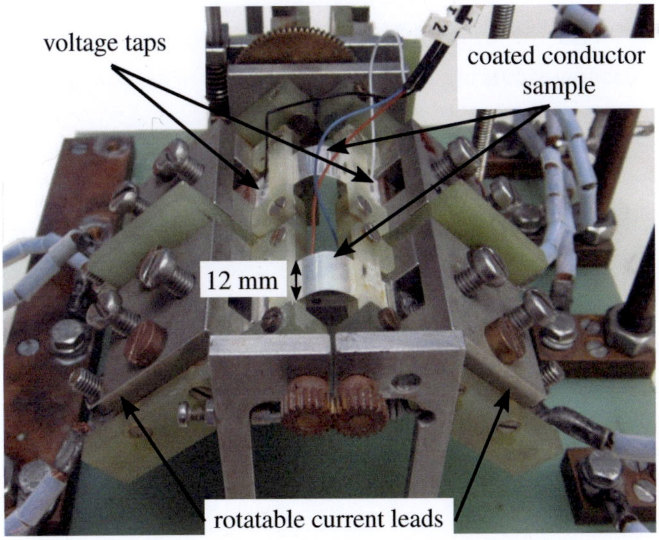

Figure 2.8: Bending machine for single tapes.

The bending angle is increased in steps of 9°. After each decrease in radius, the sample is bent back in order to distinguish between reversible and irreversible effects. The measurement sequence is therefore as follows:

$$R = \infty,\ 100.0\,\text{mm},\ \infty,\ 50.0\,\text{mm},\ \infty,\ 33.3\,\text{mm},\ \infty,\ 25.0\,\text{mm},\ \ldots$$

The critical current of the sample is determined using the electrical method: a current is applied while the voltage is measured over a 33 mm length using a nanovoltmeter. The current is increased until the criterion of 10^{-4} V/m is reached.

Since coated conductor tapes are coated with *RE*BCO on one side only, their layered structure is asymmetric, and the bending behaviour may depend on the direction of bending. Therefore, bending is tested in both directions: with the *RE*BCO layer facing inwards and outwards. For the two orientations two separate samples are used.

To compare the bending characteristics for tapes of different thickness, it is useful to estimate the strain in the *RE*BCO layer. Also, this makes it possible to compare bending data to results from different types of strain characterization such as the tensile test, Walters spring or U-spring [87, 90]. The bending strain is estimated by making several assumptions. Firstly, it is assumed that there is a neutral axis in the center of the substrate that does not change its length. The second assumption is that the bending strain in the substrate is fully determined by the shape of the substrate. These assumptions are valid as long as the superconducting layer is much thinner than the substrate and the substrate thickness is small compared to the imposed bending radius. Thirdly, the effect of silver or copper stabilization layers on the strain is neglected.

As seen in figure 2.9, the length on both sides of the substrate will differ from the neutral axis because the bending radius is not the same. This leads to a strain in the *RE*BCO layer of

$$\varepsilon = \frac{(R \pm d/2) - R)}{R} = \pm\frac{d}{2R} \tag{2.1}$$

in which d is the substrate thickness. The bending strain is negative when the *RE*BCO layer is on the inside of the bend, since it is compressed. If the *RE*BCO layer is on the outside of the bend, it is extended leading to positive strain. This value represents bending effects only and does not match the absolute strain, because thermal strain and residual strain from the production process are not taken into account.

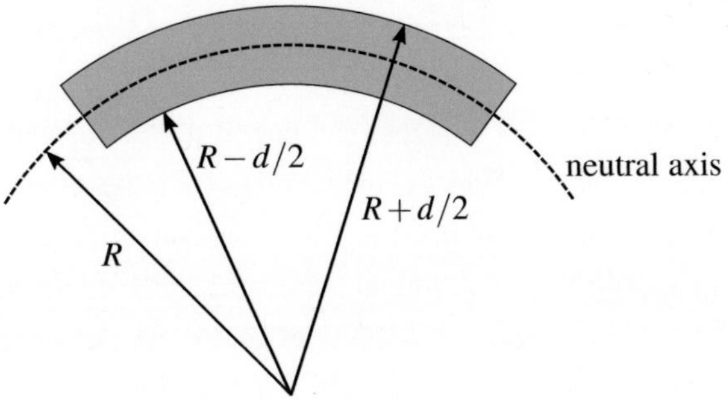

Figure 2.9: Sketch of a bent substrate of thickness d with its neutral axis. The bending strain in the superconductor can be estimated by calculating the change in length on the substrate surface compared to the neutral axis.

Materials from six different manufacturers were tested (see table 2.2). The tape from Bruker has a stainless steel substrate, while all other tapes are based on Hastelloy C-276 or a similar Ni-alloy substrate. From all manufacturers tapes with a 10 or 20 µm copper layer surrounding the tape were tested. From SuperPower, SuperOx and Fujikura, also tapes with only silver stabilizer were measured. The tapes are 12 mm wide except Fujikura FYSC-SC10, which is 10 mm wide.

Table 2.2: Investigated coated conductors and their measured bending limits. $R_{min,comp}$ and $R_{min,tens}$ are the minimum bending radii for compressive bending (*RE*BCO in) and tensile bending (*RE*BCO out). The minimum bending radius is defined as the smallest bending radius at which the irreversible degradation is less than 5%. ε_{min} and ε_{max} are the bending strains corresponding to $R_{min,comp}$ and $R_{min,tens}$ computed using equation 2.1. The used substrate material is either a Ni-alloy (N) or stainless steel (S).

manufacturer	type	substrate	stabilizer	$R_{min,comp}$	$R_{min,tens}$	ε_{min}	ε_{max}
Bruker	-	97 µm (S)	20 µm Cu	< 5.0 mm	6.7 mm	$< -1.00\%$	0.75%
Fujikura	FYSC-SC10	75 µm (N)	only Ag	< 5.0 mm	10.0 mm	$< -1.00\%$	0.40%
Fujikura	FYSC-SCH12	75 µm (N)	20 µm Cu	< 5.0 mm	8.3 mm	$< -0.75\%$	0.45%
SuNAM	HCN12500	60 µm (N)	20 µm Cu	< 5.0 mm	< 5.0 mm	$< -0.60\%$	$> 0.60\%$
SuperOx	-	60 µm (N)	20 µm Cu	< 5.0 mm	6.7 mm	$< -0.60\%$	0.45%
SuperOx	-	100 µm (N)	only Ag	< 5.0 mm	12.5 mm	$< -1.00\%$	0.40%
SuperPower	SCS12050-AP	50 µm (N)	20 µm Cu	< 5.0 mm	< 5.0 mm	$< -0.50\%$	$> 0.50\%$
SuperPower	SF12100	100 µm (N)	only Ag	5.9 mm	12.5 mm	-0.85%	0.40%
Theva	TPL 12060x	90 µm (N)	10 µm Cu	< 5.0 mm	7.7 mm	$< -0.90\%$	0.59%

The bending characteristics of all samples are shown in figure 2.10. The critical currents are normalized to the initial value, and plotted as a function of the bending strain according equation 2.1. The squares denote the critical current measured under bending strain. These measurements include both reversible and irreversible effects. The circles are the values measured after bending from back from the respective radius to $R = \infty$. These values are affected only by irreversible effects. The magnitude of the reversible strain effect can be read from the difference between the two graphs.

The minimum bending radius is defined as the lowest radius for which the irreversible degradation is less than 5%. The minimum bending radii for both bending orientations and the corresponding bending strains are shown in table 2.2. Under compressive bending, all samples but one (SuperPower SF12100) have a minimum bending radius below 5 mm, the lower limit of the bending rig. This proves again the flexibility of coated conductors under compressive bending. Most tapes are less tolerant of tensile bending, but for SuperPower SCS12050-AP and the SuNAM tape the minimum radius could not be reached either in this orientation. The lowest tolerance for tensile bending strain of 0.40% is observed in the tapes that have only a silver stabilization layer. This is in agreement with work by Shin et al. [77] and Cheggour et al. [93], who showed that the irreversible strain limit under axial tension is higher in copper-stabilized tapes.

The magnitude of the reversible effect has a large variation among the tested samples. It is rather strong in tapes from SuperPower and Bruker, while it is not observed in tapes from Fujikura. Previous works from van der Laan et al. and Sugano et al. have shown that the strain sensitivity depends on material properties including the orientation of the ab-plane [85, 94] the rare-earth metal used [84] and the level of grain alignment [81]. The variation between tapes from different manufacturers is therefore not surprising, since they employ different production techniques and tape architectures. The physical reason for the differences can however not be inferred from our measurements since those properties are unknown.

In SuperPowers's SCS12050-AP tape, a slight critical current increase of 1.7% is observed under 0.28% of tensile bending strain. In Nb_3Sn wires, such a peak occurs when the intrinsic strain becomes zero [95]. Sugano et al. found the strain at the peak in a *RE*BCO tape to shift with magnetic field, and therefore the position of the peak cannot be explained by residual strain only [96]. In the Bruker tape, a similar peak occurs at 0.20% strain. Interestingly, the samples from SuperOx, Theva and SuNAM show a peak under compression, at bending strains of −0.21%, −0.29% and −0.32% respectively.

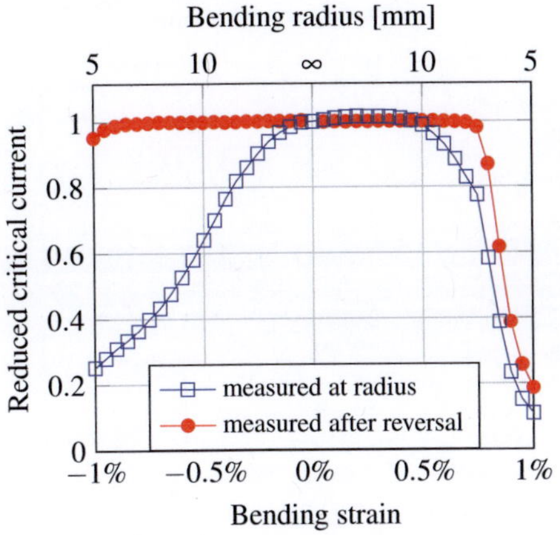

(a) Bruker tape with 97 μm thick substrate and 20 μm Cu stabilizer.

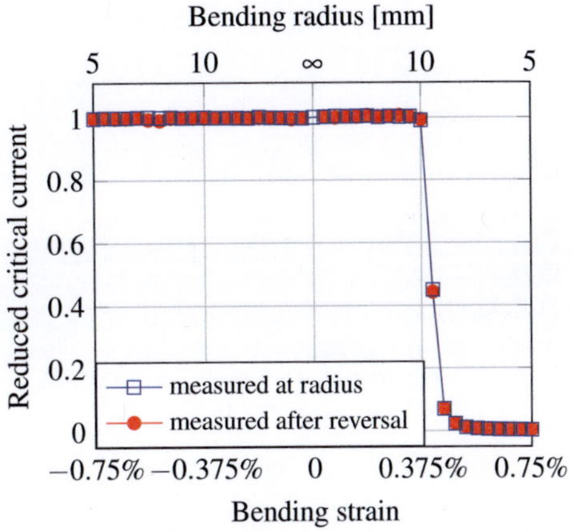

(b) Fujikura tape with 75 μm thick substrate (FYSC-SC10)

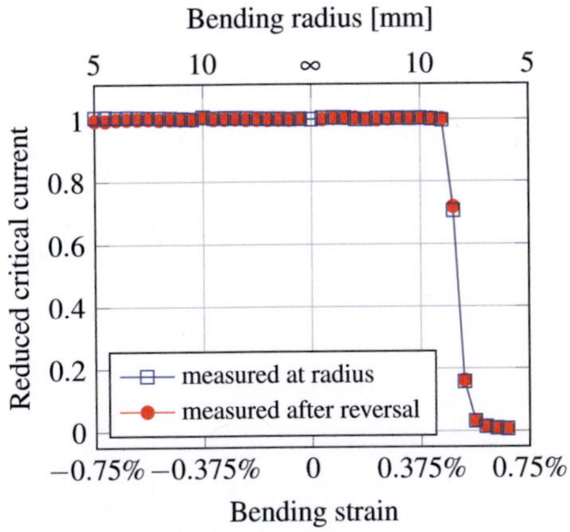

(c) Fujikura tape with 75 μm thick substrate and 20 μm of electroplated Cu stabilizer (FYSC-SCH12)

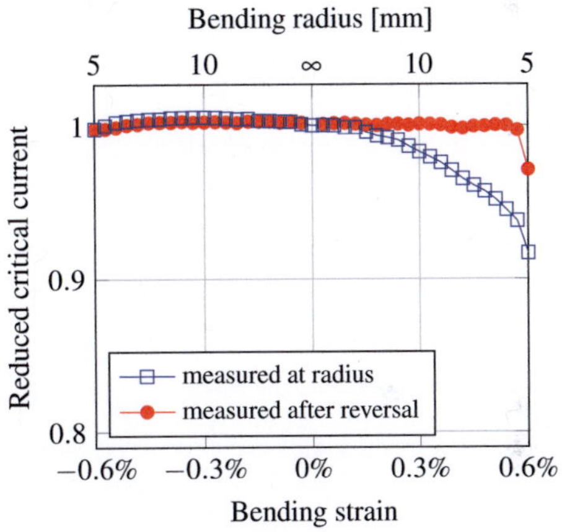

(d) SuNAM tape with 60 μm thick substrate and 20 μm Cu stabilizer (HCN12500)

Figure 2.10: Reduced critical current as a function of estimated bending strain for all measured coated conductors (continues on next pages).

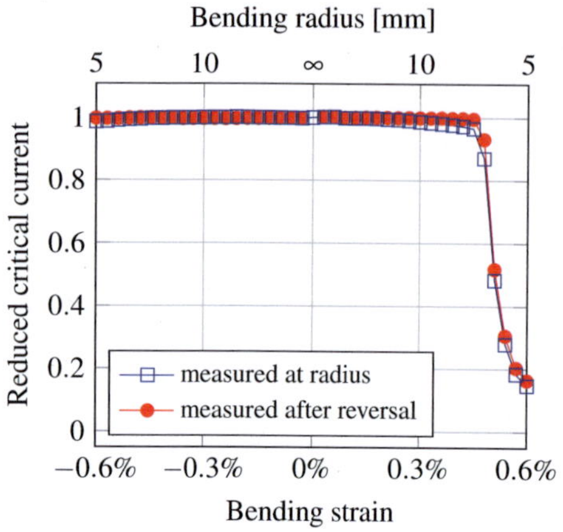

(e) SuperOx tape with 60 μm thick substrate and 20 μm Cu stabilizer

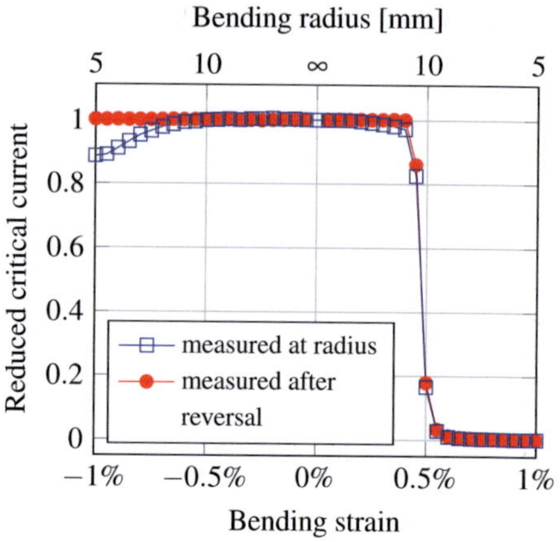

(f) SuperOx tape with 100 μm thick substrate

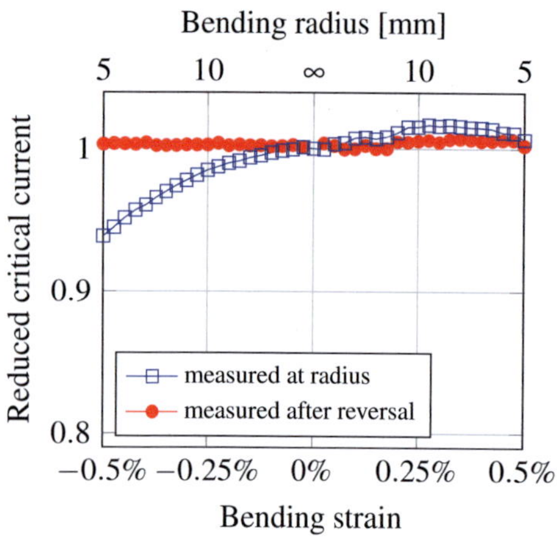

(g) SuperPower tape with 50 μm thick substrate and 20 μm Cu stabilizer (SCS12050-AP)

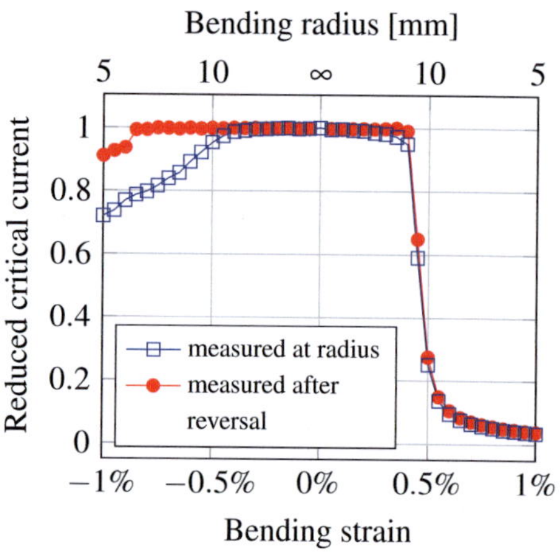

(h) SuperPower tape with 100 μm thick substrate (SF12100)

Figure 2.10: Reduced critical current as a function of estimated bending strain for all measured coated conductors (continues on the next page).

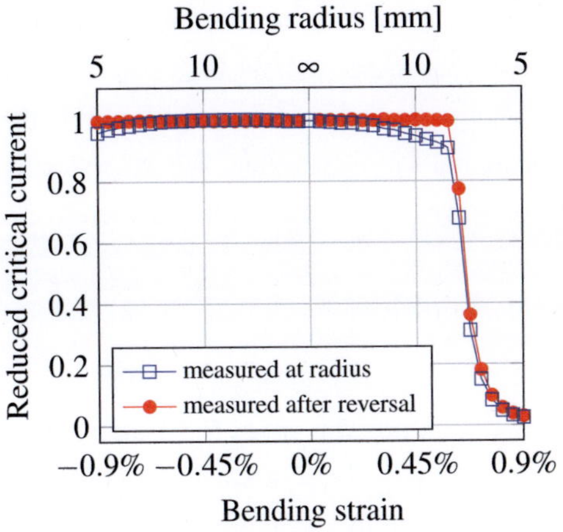

(i) Theva tape with 90 μm thick substrate

Figure 2.10: Reduced critical current as a function of estimated bending strain for all measured coated conductors.

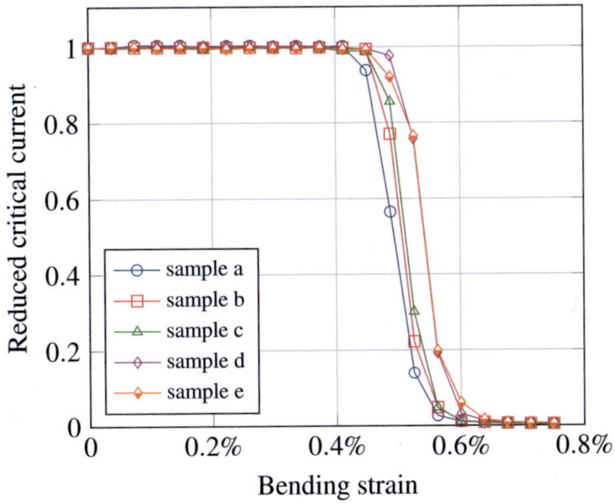

Figure 2.11: Repeated measurements for Fujikura 75 μm thick substrate and 20 μm of electroplated Cu stabilizer (FYSC-SCH12). Five samples named 'a' to 'e' were taken from the same tape and bent in tensile direction in order to check for variation in measured bending characteristic. The bending strain limits for these five sample differ by no more than 0.1%.

In figure 2.11, measurements are shown for five samples taken from the same tape (Fujikura FYSC-SCH12). It is evident that there is about 0.1% variation in measured bending strain limit in the same tape. This could result from inhomogeneities in the coated conductor or from experiment conditions that are not fully repeatable. When applications are close to the bending strain limit, bending tests of longer length (test coils) are therefore recommended.

2.3.3 Bending of Roebel cables

In addition to the single tapes, bending tests were performed on two Roebel cables. The strands in such a cable are free to slide along the cable for some distance depending on the cable architecture. Therefore, the strands do not experience additional bending strain from the cable assembly, and a bending behaviour similar to single tapes is expected.

To characterize the bending properties of Roebel cables, a new and larger bending device was constructed (figures 2.12 and 2.13). The device can accommodate cable samples longer than one transposition length and larger current contacts. Unlike the tape bending machine, the cable sample has a constant bending angle of 180°. The radius of this 180° bend can be decreased by screwing together two blocks of glass-fibre reinforced plastic (GFRP). A consequence of this method is that the length of the bent cable segment ℓ is not constant, but depends on the radius:

$$\ell = \pi R \tag{2.2}$$

To compensate for this change in length, the current leads are moved under an angle, as illustrated in figure 2.12. If the radius decreases by ΔR, the length of the bent segment decreases by $\Delta\ell = \pi\Delta R$. To compensate for this, each current contact needs to move to the right by $\Delta\ell/2$. The desired angle for the current lead track is therefore:

$$\tan(\gamma) = \frac{\Delta R}{\Delta\ell/2} = \frac{2}{\pi} \quad \Rightarrow \quad \gamma \approx 32.5^\circ \tag{2.3}$$

The cable can be bent only with decreasing radius; reversed bending must be done manually at room temperature. The radius of the cable is defined as the distance between the moving GFRP blocks divided by two. This equals the outer bending radius of the cable. Strands located more to the inside of the cable have a slightly lower bending radius.

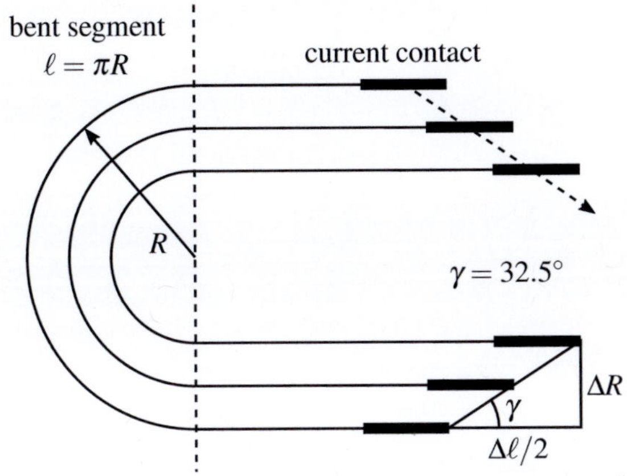

Figure 2.12: The sample shape at different radii. The current contacts move under and angle $\gamma = 32.5°$ in order to compensate for the change in length in the bent segment. The bent segment keeps the half-circle shape at all times.

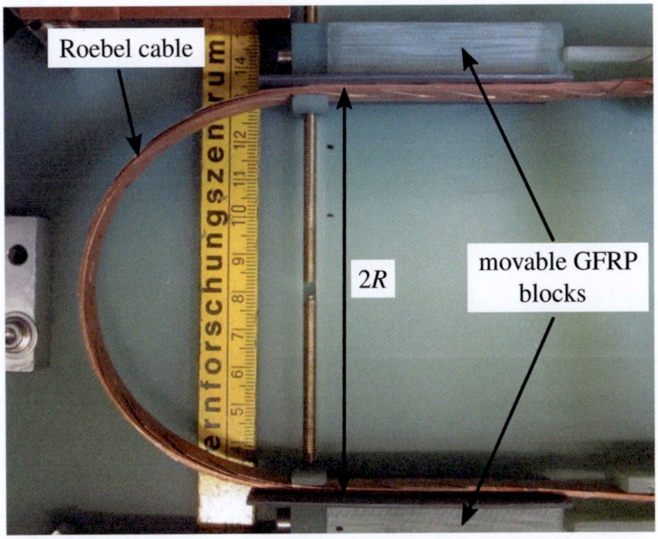

Figure 2.13: Picture of the bent segment of the cable on the bending device. The radius of the bend can be decreased by screwing together two GFRP blocks.

Two 105 cm long Roebel cables were prepared: one from SuperPower (SCS12050-AP) and one from a tape from Bruker (see table 2.2). Both cables had a transposition length of 226 mm and consisted of 15 strands. Cross-sections of both cable are shown in figure 2.14. Both cables are assembled in such a way, that the *RE*BCO layer of the strands are facing the substrate of the next tape.

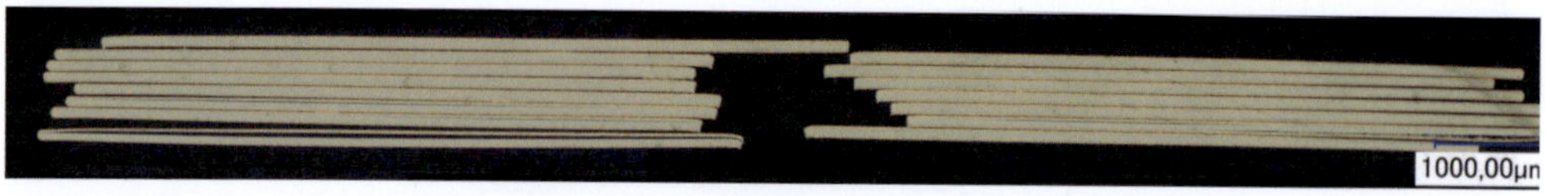

(a) Roebel cable from SuperPower SCS12050-AP

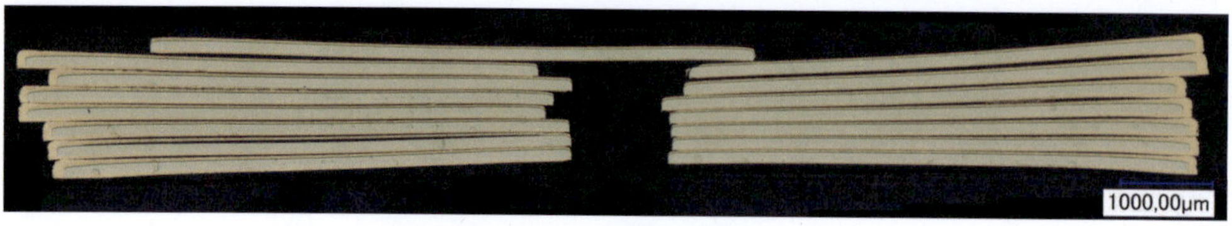

(b) Roebel cable from Bruker coated conductors

Figure 2.14: Cross-sections of the Roebel cables.

Both SuperPower and Bruker add copper to the coated conductor by electrodeposition. The high aspect ratio of the tape makes it difficult to reach a homogeneous current density in the electrolyte. This can lead to an excess of copper deposited at the tape edges. This "dog-bone" is especially visible in the tape from Bruker. Since the tapes in a Roebel cable are stacked, the increased thickness at the edges is also noticeable in the cable cross-section.

The critical currents at 77 K of the separate strands were measured before the cables were assembled. Current terminals were soldered with Sn60Pb40 over one transposition length to each end of the cables. To each strand, a pair of voltage tapes was soldered over a length of 45.2 cm, which equals two transpositions. The critical current is determined using a criterion of $E_c = 1$ μV/cm. The cable critical current is then computed by averaging the values of each strand.

The individual strands had an average critical current of 151.6 A and 58.9 A for the SuperPower and Bruker cable respectively. After assembly, the critical current of the cable was first measured at a radius of 50 mm in compressive bending. The initial critical currents were 1414 A for the SuperPower cable and 658 A for the Bruker cable. Figure 2.15 shows the critical currents as a function of the outer bending radius, normalized to the value at 50 mm. For comparison, the results of the single tape bending of SuperPower and Bruker material are also shown.

The SuperPower cable was bent to an outer radius of 4.0 mm, at which the critical current had decreased by 5.9%. This decrease was found to be fully reversible after a final measurement at 50 mm radius. The Bruker cable was bent to 9.4 mm, decreasing the critical current by 35.7%. As in the SuperPower cable, the degradation was fully reversible. Such a very strong reversible effect was also noted in single tape bending with tapes from Bruker (see figure 2.10a). The importance of this reversible decrease is limited, because the Bruker material is intended for use in high-field magnets that operate at 4.2 K. The reversible strain effect at this temperature is weaker than at 77 K [80, 90]. The effect however needs to be taken into account if the magnet is pre-tested at 77 K.

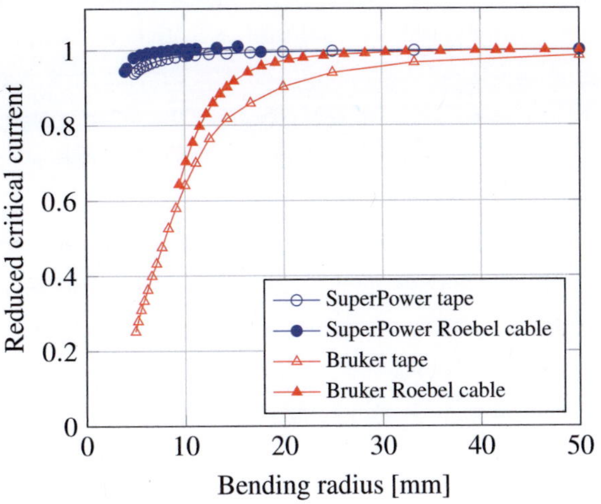

Figure 2.15: Critical current as a function of bending radius for the Roebel cables from SuperPower and Bruker coated conductors. The single tape results are shown for comparison. The *RE*BCO layer is facing inwards in measurements of both tapes and cables (compressive bending). The measurement was done at 77 K and in magnetic self-field.

Both cables show a weaker bending radius dependence than their single tapes. This is likely caused by the different nature of the bending experiments: in the single tape bending machine (figure 2.8), the length of the bent segment is constant, while in the cable bending machine (figure 2.13), it becomes shorter with decreasing radius. Strain effects may therefore be more pronounced in the single tape experiments. Secondly, the reversible strain effect depends on the magnetic field and can be weakened by small magnetic fields in the range of 0.2 – 0.6 T [80]. Since the conductors experience a higher self-field in Roebel cables than in single tape measurements, this may affect the bending properties.

2.3.4 Conclusion

The out-of-plane bending characteristics were investigated at 77 K in coated conductors from SuperPower, Bruker, SuperOx, Fujikura, Theva and SuNAM. The measurements confirm the high out-of-plane bending flexibility of coated conductors: all samples could be bent to 6 mm radius in compressive mode with less than 5% permanent degradation.

There is, however, a large variation in bending properties, which cannot be explained only by a difference in substrate thickness. The conductors from SuperPower, SuperOx, Fujikura and Theva are more tolerant of compressive bending than tensile bending. The Bruker tapes show a strong reversible effect, while in tapes from SuperPower, SuperOx, Theva and SuNAM this effect is much weaker, and in tapes from Fujikura it is not observed. The physical reason for these differences cannot be inferred from the measurements.

Roebel cables made from SuperPower and Bruker coated conductors were also examined. The cables were bent down 4.0 and 9.4 mm respectively, without irreversible degradation. The Roebel cable assembly does not seem to severely limit the out-of-plane bending properties, and therefore such cables are suitable for low-radius windings such as an aligned block coil or cos-theta coil, which have minimum bending radii of 16 and 7.5 mm. Both cables showed a reversible decrease in critical current of up to 5.9% (SuperPower) and 35.7% (Bruker) at the lowest measured radius. Although no damage conductor is observed, this effect still may influence the coil performance. The reversible strain effect depends on the temperature and the magnetic field and therefore needs to be investigated in the conditions of the specific application. Additionally, the effects of *in*-plane bending and cyclic loading on Roebel cables remain to be investigated.

2.4 Mechanical test of the coil ends of the CEA cos-theta design

A critical aspect in CEA's cos-theta magnet design [49] is the bend the cable needs to make to exit the coil from the inner winding (figure 2.16). Here the cable makes an S-shaped bend and undergoes a twist of 74° over a length of 80 mm, which equals a twist pitch of 389 mm. This section describes a test that was done to investigate the influence of such deformation on the cable.

In order to test the effect of such deformation on the Roebel cables, plastic moulds resembling the coil exit were made. The moulds were designed at CEA and 3D-printed in Bluestone resin[2] at CERN. The cable can be fixed in such a mould to impose a specific bending radius and twist pitch. Three moulds with increasing severity of deformation were made (table 2.3). Mould 1 has the same shape as the actual coil exit, while mould 2 and 3 are intermediate steps with less severe deformation. Mould 2 has no S-bend and mould 3 has an increased twist length of 110 mm.

Roebel cables can be laid assembled as a left-hand or right-hand turning helix. Since the moulds have a left-hand twist, a left-hand Roebel cable would be overtwisted, while a right-hand one would be untwisted. The effect of over- and undertwisting has not been investigated and the

[2] Accura Bluestone: `https://www.3dsystems.com/materials/accura-bluestone`

cable lay for the magnet design is not specified. For this reason, both a left-hand (cable 1) and a right-hand Roebel cable (cable 2) were prepared. The cables were prepared from 12 mm wide coated conductor from SuperPower (SCS12050-AP). The properties of this conductor are listed in table 2.2 in the previous section. The cables consisted of 15 strands and had a transposition length of 300 mm. The transposition length is different from the one used in the transverse stress and bending experiments because the EuCARD-2 cable specifications changed in the meantime.

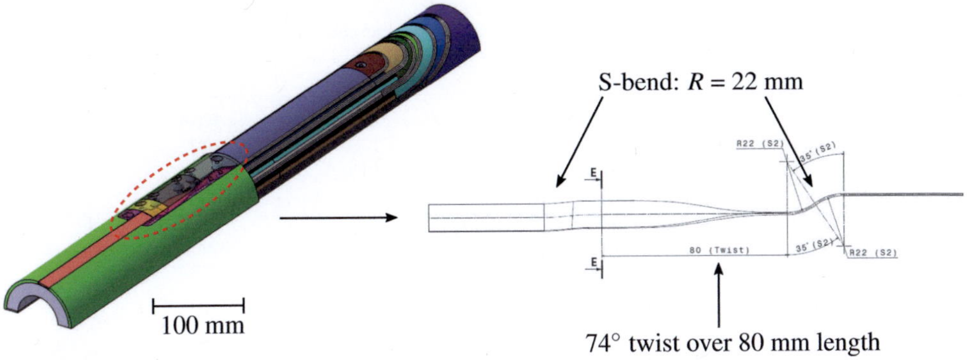

Figure 2.16: The cos-theta magnet design and the cable shape at the coil exit. Here the cable makes an S-bend with a radius of 22 mm and has a 74° twist over a length of 80 mm (389 mm twist pitch). Images owned by C. Lorin and M. Durante (CEA).

Table 2.3: Torsional twist pitch and bending radius of the moulds. Mould 1 has the same deformation as the actual coil exit. Mould 2 and 3 are intermediate steps with less severe deformation.

mould	twist angle	twist length	twist pitch	bending radius
no. 1	74°	80 mm	389 mm	22 mm
no. 2	74°	80 mm	389 mm	no bend
no. 3	74°	110 mm	535 mm	no bend

The critical current was first measured without a mould, then in moulds with increasing level of deformation. All measurements were done in a liquid nitrogen bath and in magnetic self-field. The results are shown in figure 2.18. Both cables had a slightly increased critical current when mounted in the mould. Cable 2 was measured once more after removal from the mould. The critical current reverted to the original value, indicating that the increase was reversible. The reason for the small increase is not yet understood.

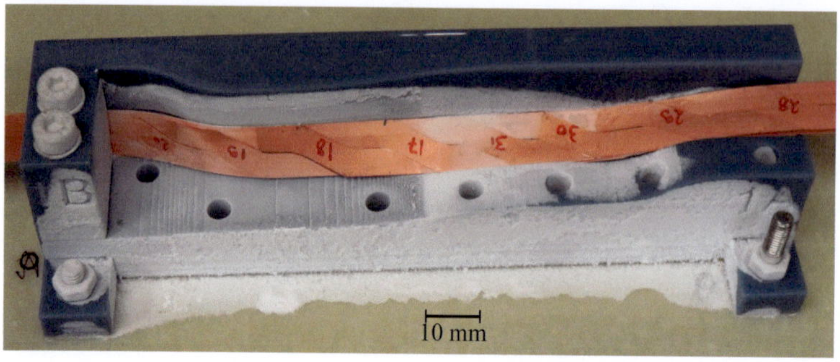

Figure 2.17: Mould no. 1 with the right-hand assembled Roebel cable (cable 2) after being opened.

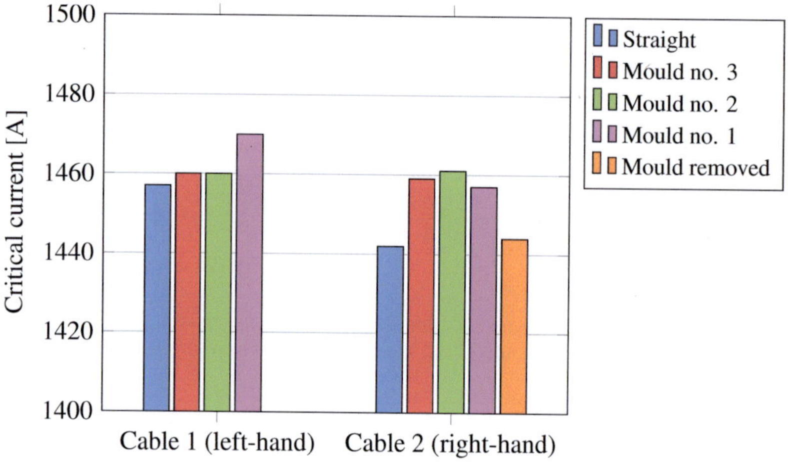

Figure 2.18: Measured critical currents at 77 K of both cables for all deformation steps. The measurements are shown in chronological order.

Two short Roebel cables were subjected to the deformation that occurs at the coil exit of CEA's cos-theta coil. Both cables withstood deformation without degradation of the critical current. The design of the coil ends can therefore be validated.

Chapter 3

Inter-strand resistance

3.1 Introduction

The resistance between strands in a superconducting cable influences several aspects of the cable performance. On the one hand, a sufficiently low resistance allows current redistribution between strands. This can potentially improve stability of the cable, since current can redistribute and “flow around" a defect or normal zone in one strand. On the other hand, a low resistance can lead to an increase in AC losses due to induced coupling currents between the strands.

In figure 3.1, a cross-sectional sketch of a Roebel cable is shown. Several properties of the inter-strand connections in Roebel cables can be derived from this perspective. Firstly, strands are connected only to their neighbours, because the cable has an empty core. Secondly, the full transposition means that each strand is equivalent, and that the resistance between each pair of neighbouring strands is the same. These features make the inter-strand connections in a Roebel cable similar to those in a Rutherford cable with an insulating core. Thirdly, all connections are essentially front-to-back contacts of two coated conductors, because the superconducting layer in each strand faces in the same direction.

This chapter aims to find a description of inter-strand resistances in Roebel cables. In section 3.2, a formula is derived that expresses the contact area between strands in terms of the cable parameters. A new set-up was built to investigate resistance of two-tape contacts and inter-strand resistance in short Roebel cables as a function of applied pressure. This set-up is described in section 3.3. The two-tape contacts were investigated because the interpretation of results is easier than in more complicated structures such as Roebel cables. These are discussed in section 3.4. A two-parameter model to describe inter-strand connections in Roebel cables is proposed in section 3.5. This model is based on an existing model for Rutherford cables, which have the same helical transposition of strands. The results for Roebel cables are compared to those of two-tape contact by normalizing the inter-strand resistance to the contact area between strands. The results of this chapter are used as an input to the calculations of AC loss and current redistribution, which are described in later chapters 4 and 5 .

Figure 3.1: Sketch of a cross-section of an assembled Roebel cable with 10 strands. The yellow dashed lines illustrate the "front-to-back" contacts between the strands. Because of the empty core of the cable, strands are connected only to their neighbours.

3.2 Contact areas between strands

In this section, a formula for the theoretical contact area between strands in a Roebel cable is derived. This is done by calculating the overlap area of two adjacent strands extracted from a cable. Figure 3.2 shows a sketch of two such strands. The strands are shifted along the cable axis by ℓ_t/N, where ℓ_t is the transposition length and N is the number of strands. The contact width varies along the cable axis x: between x_0 and x_1, the strands are in contact over their entire width w_s, while between x_2 and x_3 there is no contact. This pattern repeats along the cable axis after each half transposition $\ell_t/2$ (see figure 3.3).

The values of x_1, x_2, x_3 and x_4 can be found from figure 3.2 using simple geometrical constructions:

$$x_1 = x_0 + \frac{w_c}{2\sin(\alpha)} \tag{3.1}$$

$$x_2 = x_1 + \frac{w_s}{\tan(\alpha)} = x_0 + \frac{w_c}{2\sin(\alpha)} + \frac{w_s}{\tan(\alpha)} \tag{3.2}$$

$$x_4 = x_0 + \frac{W}{\tan(\alpha)} + \frac{\ell_t}{N} - \frac{w_c}{2\sin(\alpha)} \tag{3.3}$$

$$x_3 = x_4 - \frac{w_s}{\tan(\alpha)} = x_0 + \frac{W - w_s}{\tan(\alpha)} + \frac{\ell_t}{N} - \frac{w_c}{2\sin(\alpha)} \tag{3.4}$$

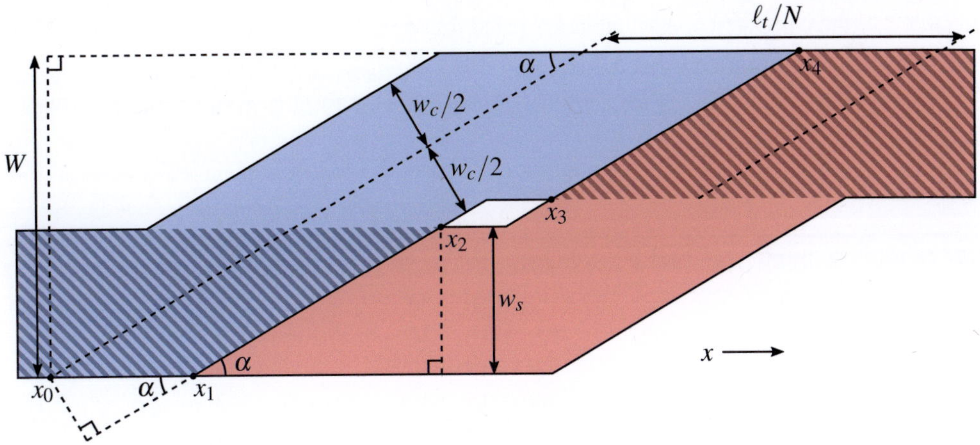

Figure 3.2: Sketch of two adjacent strands extracted from a Roebel cable. The striped area denotes the contact area between the two strands. The inner radius is of the strands is assumed to be zero.

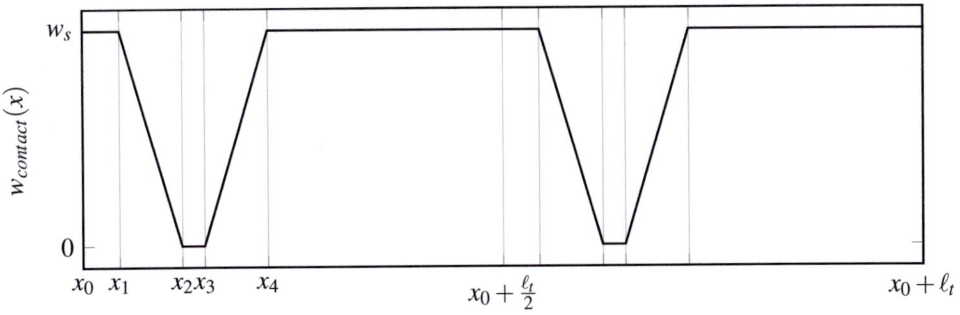

Figure 3.3: Width of the contact surface between two adjacent strands over one transposition length. The pattern repeats after each half transposition $\ell_t/2$.

The contact area per transposition length can be found by calculating the area under the graph in figure 3.3:

$$A_{contact} = 2\left((x_1 - x_0)w_s + \frac{1}{2}(x_2 - x_1)w_s + \frac{1}{2}(x_4 - x_3)w_s + (x_0 + \frac{\ell_t}{2} - x_4)w_s\right) \tag{3.5}$$

By filling in equations 3.1-3.4 one finds:

$$A_{contact} = 2w_s\left(\ell_t\left(\frac{1}{2} - \frac{1}{N}\right) + \frac{w_c}{\sin(\alpha)} - \frac{W - w_s}{\tan(\alpha)}\right) \tag{3.6}$$

In table 3.1, the computed contact areas are listed for several common cable architectures. The average width of the inter-strand contact is defined as $w_{contact,av} = A_{contact}/\ell_t$. For the cable architectures shown, average contact width is in the range of 79-88% of the strand width w_s.

Table 3.1: Inter-strand contact area per transposition length and average contact width for several common cable architectures.

W [mm]	w_s [mm]	w_c [mm]	ℓ_t [mm]	N	α	$A_{contact}$ [mm^2]	$w_{contact,av}$ [mm]
4	1.9	1.9	116	10	30°	177	1.53
12	5.5	5.5	126	10	30°	552	4.38
12	5.5	5.5	226	15	30°	1074	4.75
12	5.9	5.9	300	15	30°	1547	5.16
12	5.9	5.9	300	13	30°	1512	5.04

3.3 Press set-up for measurements in liquid nitrogen

The strands in a Roebel cable are loosely fitting and able to move. As a result, no well-defined contact exists between strands when no pressure is applied. An external pressure therefore needs to be applied for proper measurements of inter-strand resistance. For this purpose, a special press was constructed to apply a transverse force to the sample in a liquid nitrogen bath (see figure 3.4). The force is generated by a hydraulic actuator. Although the actuator is capable of 200 kN, the maximum force is limited to 40 kN by the rest of the construction. The force is transferred to the sample by two glass-fibre reinforced plastic (GFRP) anvils. The length of the upper anvil is 126 mm, which equals one transposition length of the tested Roebel cables.

The force is measured by two custom-made load cells built in the vertical parts of the press. The load cells consist of CuBe cylinders with four strain gauges in a Wheatstone bridge configuration. When supplied with a constant current, such configuration ideally results in a linear response of voltage to applied force. The load cells were calibrated in the ITEP CryoMAK [97] tensile facility at 77 K and with 1.00 mA of applied current. The results of of the calibration are shown in figure 3.5 and table 3.2. Although the load cells have a different offset, the calibration factor is similar and the response to force is linear.

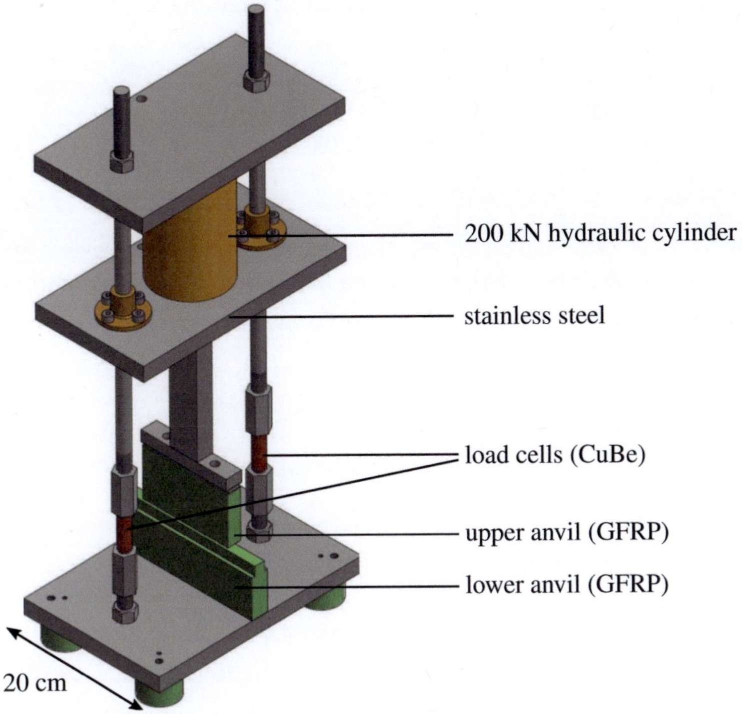

Figure 3.4: Press set-up for use in a liquid nitrogen bath.

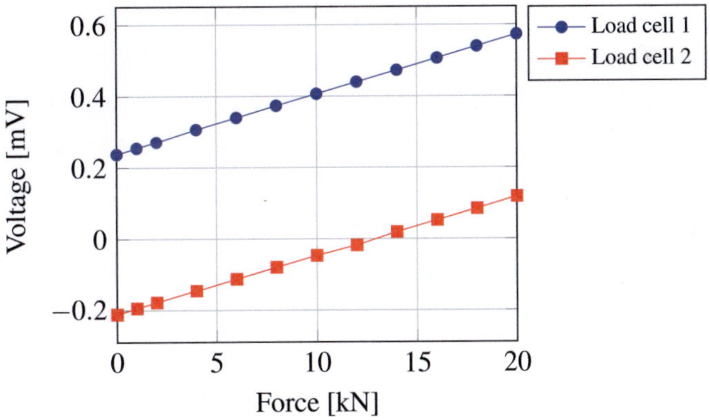

Figure 3.5: Calibration of the load cells at 77 K with 1.00 mA current applied.

Table 3.2: Calibration factor and offset for both load cells obtained from the data in figure 3.5.

	calibration factor [mV/kN]	offset [mV]
load cell 1	0.01670	0.23948
load cell 2	0.01632	-0.20920

The weight of the press above the load cells makes a considerable contribution to the sample force that is not measured by the load cells. The mass above the load cells is 28.1 kg resulting in 275.7 N downwards force. The total sample force is therefore given by:

$$F_{\text{sample}} = 275.7\ \text{N} + F_{\text{load cell 1}} + F_{\text{load cell 2}} \tag{3.7}$$

where $F_{\text{load cell 1}}$ and $F_{\text{load cell 2}}$ are computed using the calibration values.

3.4 Resistance of two-tape contacts

This section describes resistance measurements of two-tape contacts. Such contacts in front-to-back orientation are the building blocks of inter-strand contacts within a Roebel cable. The simple geometry makes a two-tape contact easier to prepare than a Roebel cable. It also simplifies the interpretation of measurement data.

Many investigations of contacts of two coated conductor tapes are available in literature. Most of these investigations, however, aim to develop joints with a low resistance. The tapes in these joints are usually in front-to-front orientation, and the resistance is reduced by techniques as soldering [98], inserting indium foils [99, 100], or surface treatments to remove copper oxide [101, 102]. Such contacts are not representative for the front-to-back contacts that occur in standard Roebel cables. Data on soldered front-to-back contacts is only sparsely available [103–105], and for pressed front-to-back contacts no data could be found.

The two-tape contacts investigated in this section are not made to achieve the lowest resistance, but to resemble the behaviour of contacts within a Roebel cable. The surface preparation is kept similar to that of the Roebel cables which are described in section 3.5. Copper surfaces are not polished or etched to remove oxide layers.

3.4.1 Sample preparation

The contacts were prepared from two 80 mm long pieces of 12 mm wide coated conductor. The surface was cleaned using acetone to remove protective polymer layers that manufacturers sometimes apply to the surface[1]. Polishing the surface removes copper oxide layers and reduces roughness, and would likely lead to a lower contact resistance. Yet we chose not to polish the surface so that the measurements are representative for our Roebel cables, of which the strands are also not polished. The two tapes are mounted on a sample holder with 40 mm overlap (see figure 3.6). A force is applied to the overlapping area using a GFRP pressure anvil. A current of 10 A is applied through the contact. The voltage between the two tapes is measured using a nanovoltmeter. Since 10 A is far below the critical current of more than 300 A, the potential difference within the superconductor can assumed to be zero. The measured voltage can therefore be fully ascribed to the contact resistance.

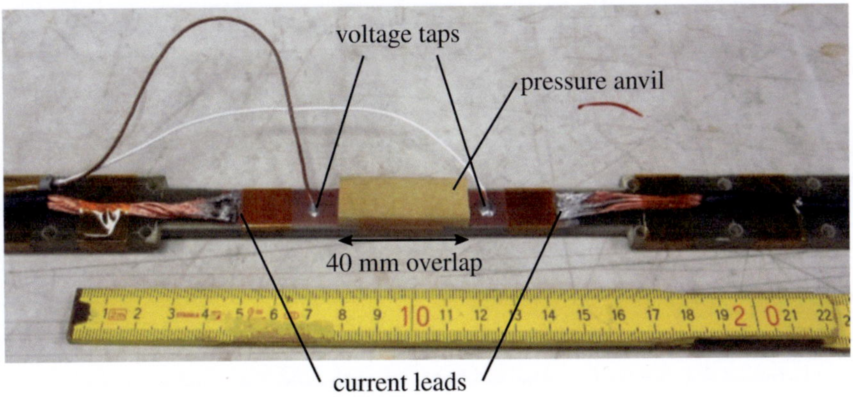

Figure 3.6: The sample holder with a two-tape press contact on it.

Besides press contacts described above, several soldered contacts were measured as well. Soldered contacts of SuperPower tapes were prepared as follows: first, the samples are cleaned with acetone and a liquid rosin-based flux[2] is applied to the surface. The tapes are then pre-soldered with In52Sn48 at 200 °C on a hot plate within one minute. Next, the tapes are joined together with 40 mm overlap using the purpose-made tool shown in figure 3.7. By compressing six springs by 2.5 mm, a force of about 940 N is applied. This equals an average stress of 2.0 MPa. The tool is then heated in an oven to 170°C for 90 minutes to melt the solder once more. The tool with the sample is left to cool to room temperature before the force is released.

Soldered contacts from SuperOx tapes were already coated with solder by SuperOx. They were directly joined using the soldering tool without pre-soldering.

[1] Personal communication with A. Molodyk (SuperOx)

[2] Interflux IF 6000 `http://interflux.com`

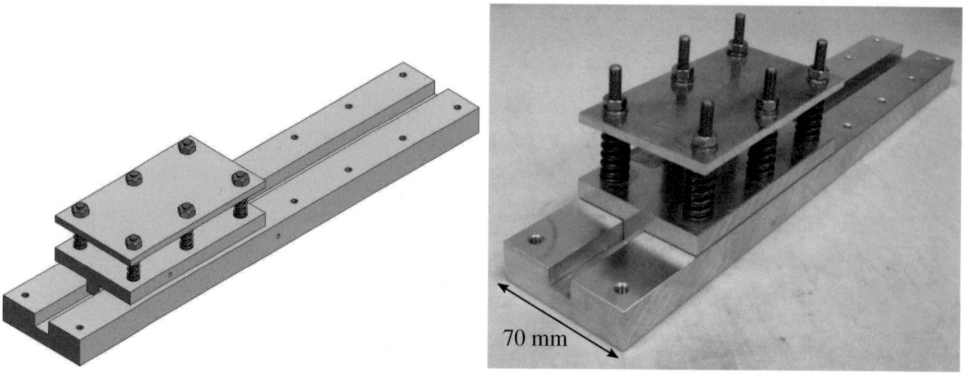

(a) Computer drawing **(b)** Picture

Figure 3.7: Aluminium tool for soldering tapes and cables with an applied force.

3.4.2 Results: SuperPower coated conductors

Pressed and soldered contacts were prepared from SuperPower SCS12050-AP coated conductor tape. This tape is 12 mm wide, has a 50 μm thick Hastelloy C-276 substrate and is coated with 20 μm of copper surrounding the tape. Contacts were prepared with the *RE*BCO coated sides facing each other ("front-to-front") and facing the same direction ("front-to-back"). The latter is the type of contact that occurs in a Roebel cable. Of each type of contacts at least three samples were prepared in order to test the repeatability.

The measured contact resistivities as a function of transverse stress are shown in figure 3.8. Both the resistance and the applied force are normalized to the contact area of 40 mm × 12 mm. The soldered front-to-front contacts have a surface resistivity in the range $1.3 - 2.2 \cdot 10^{-11}$ Ωm^2. This is in the same order of magnitude as resistivities found in soldered splices [103–105], although those splices were not prepared with the same procedure and the splice resistance of SuperPower wires was found to depend on the batch number [103]. The resistivities of front-to-back soldered contacts are two orders of magnitude higher with $1.0 - 1.3 \cdot 10^{-9}$ Ωm^2, because of the highly resistive substrate and buffer layers. The large difference between front-to-front and front-to-back contacts means that the front-to-back internal resistance dominates over the interfacial resistance between layers and resistance of the solder layer. The latter two parameters will therefore not have a large influence on the inter-strand resistance in Roebel cables.

Unlike the soldered contacts, the contacts that were only pressed show a clear pressure dependency. The resistance of multiple press contacts prepared in the same way was not repeatable; resistivities are spread over more than an order of magnitude. The exact reason for this sample-to-sample variation is not understood. A likely cause, however, is the variation

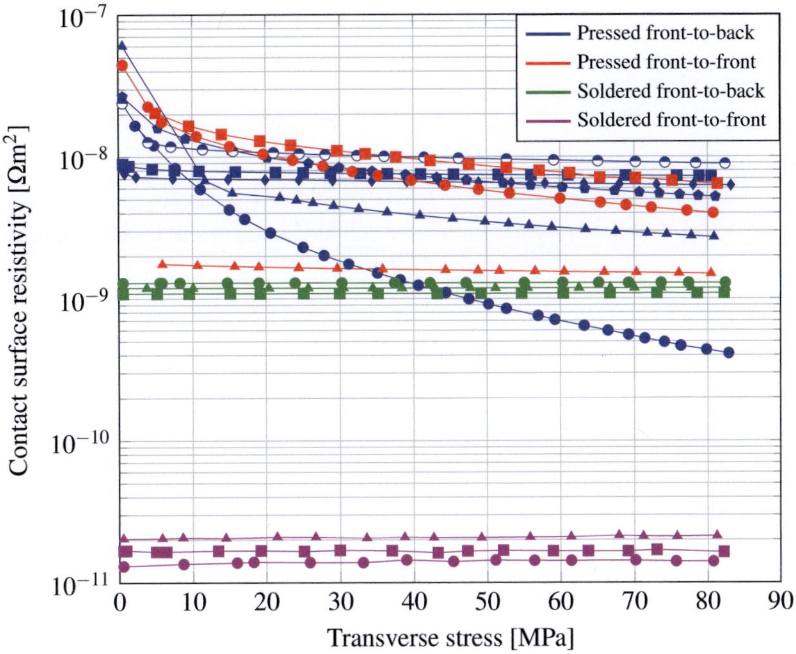

Figure 3.8: Contact surface resistivity as function of transverse stress for different two-tape contacts prepared using SuperPower coated conductor. Graphs with the same color represent different samples prepared in the same way.

in thickness of the copper layer along the length and width of the tape. Most manufactures of coated conductors add copper by electro-deposition. The deposition rate is determined by the current through the electrolyte bath. In wide tapes with a high aspect ratio this leads to an excess of copper deposition on the edges, which is called "dog-boning". Some manufacturers use techniques to suppress dog-boning. For example, Bruker shields the edges of the tape against excess electrolytic current [31]. Some thickness variations can however persist despite such corrections. Variation in thickness up to 20 μm over the tape width have been observed in SuperPower tapes [86]. When two such tapes with inhomogeneous thickness are pressed together, the actual contact area is only a fraction of the total surface area. The size of this area cannot be controlled, and is a probable reason for the large variation between samples. This problem can be overcome by inserting a layer of indium between the two tapes [106]. Indium is a relatively soft metal and, under pressure, compensates for the variation in copper thickness. This method is however not practical for Roebel cables. The important message of these observations is that a large variations of inter-strand resistance can also be expected in Roebel cables that are merely pressed.

No clear difference between front-to-back and front-to-front pressed contacts can be observed because the sample-to-sample variation is larger than the internal resistance of about $10^{-9}\ \Omega m^2$.

3.4.3 Results: SuperOx coated conductors

From SuperOx tapes with different amounts of copper were available (5, 10 and 20 μm), all based on a 60 μm thick Hastelloy C-276 substrate. From these tapes front-to-back press contacts with different copper layers were prepared. The measured resistivities are shown in figure 3.9. In addition, solder coated tapes with InSn and InAg3 were available. From these tapes, soldered front-to-back contacts were prepared, of which the resistivities are also shown in figure 3.9. The soldered coated tapes have a 60 μm thick substrate and 20 μm of copper stabilization.

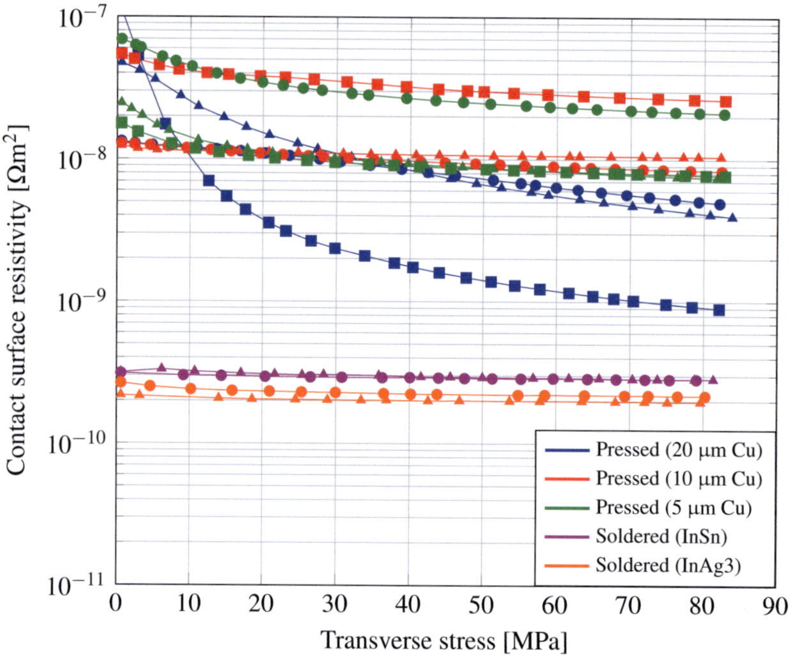

Figure 3.9: Contact surface resistivity as function of transverse stress for different two-tape contacts made from SuperOx tapes. In all samples, the superconducting layers face the same direction (front-to-back). Graphs with the same color represent different samples prepared in the same way.

Similar to the SuperPower tapes, the press contacts show a large sample-to-sample spread. There is no clear difference between contacts with 5 or 10 μm of copper. The contacts with 20 μm copper had on average a lower resistance and a stronger pressure dependency. This is possibly the results of a larger variation in copper thickness and a smaller internal resistance due to an increased amount of copper on the tape edges. All soldered front-to-back contacts had a pressure independent resistance somewhat lower than those observed in soldered contacts from SuperPower tapes.

3.5 Inter-strand resistance in Roebel cables

3.5.1 Network model and measurement method

Roebel cables are topologically equivalent to Rutherford cables (see figure 3.10). For this reason, the inter-strand resistance measurements described in this chapter are done using an established method for Rutherford cables. In Rutherford cables, inter-strand contacts are commonly described using a network model, which is shown in figure 3.11. This model has two inter-strand resistance parameters: R_a, the elementary resistance between adjacent strands, and R_c the resistance at cross-overs of strands from the upper and lower layer. Adjacent connections of R_a exist only between neighbouring strands, while crossing connections of R_c connect any pair of strands. The parameter R_a is defined such that it occurs twice in each cell of length ℓ_t/N. Cross-over connections of R_c occur twice per transposition length for each pair of strands. This network has been used in many works including those by Sytnikov et al. [107] and Verweij et al. [108] to predict AC coupling losses in Rutherford cables.

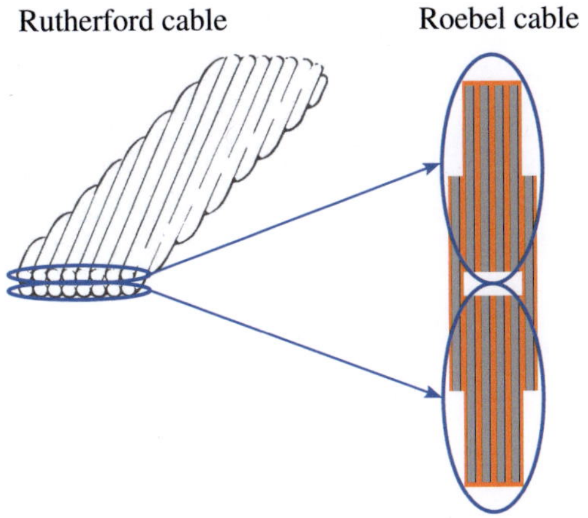

Figure 3.10: *RE*BCO Roebel cables are topologically equivalent to Rutherford cables of round wire. The narrow side a Rutherford cable becomes the wide face of a Roebel cable when the round strands are replaced by tapes with a high aspect ratio. Rutherford cable image by Wilson [109, p. 308].

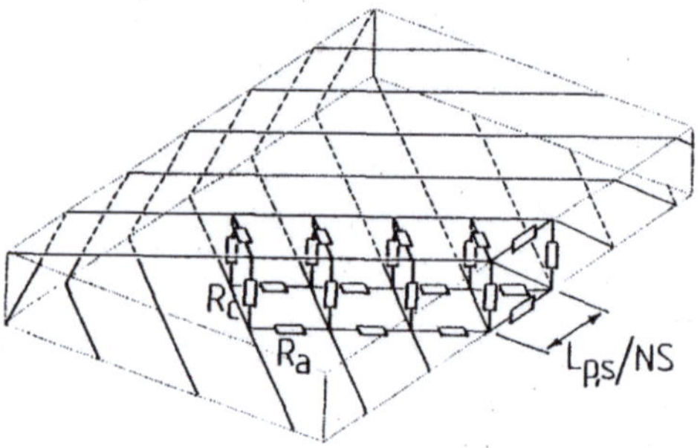

Figure 3.11: Network model of a Rutherford cable used by Verweij et al. [108].

In Rutherford cables, R_c corresponds to the resistance of the contact where a strand of the lower layer touches a strand in the upper layer. Since no such localized contact points exist in Roebel cables, it makes more sense to use parameters that are normalized to unit-length. For this the inter-strand resistance parameters ρ_a and ρ_c are introduced: ρ_c is the unit-length resistance connecting any pair of strands, and ρ_a is the unit-length resistance connecting adjacent strands. Both parameters have units of Ωm. The relation to the Rutherford cable model is as follows:

$$\rho_a = \frac{\ell_t}{2N} R_a \tag{3.8}$$

$$\rho_c = \frac{\ell_t}{2} R_c \tag{3.9}$$

To find the values of ρ_a and ρ_c, the same method is used as in a work by Devred et al. [110]. A current in passed between two strands which are opposite in the cable while the potential is measured on each strand. An electrical scheme of this set-up is shown in figure 3.12. The resulting voltage profile is then fitted to a profile calculated from the network model by using ρ_a and ρ_c as fitting parameters.

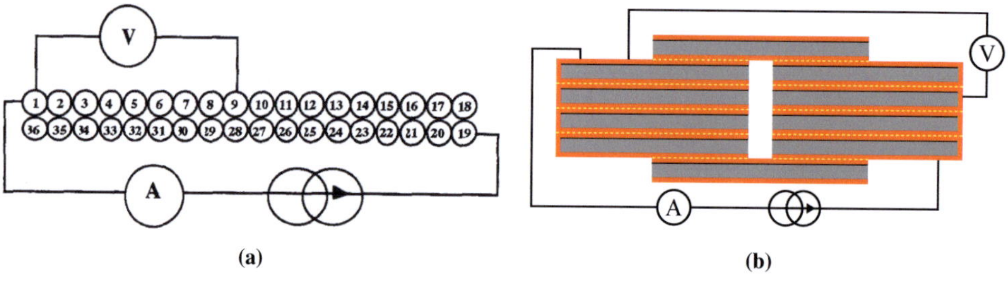

Figure 3.12: (a) Instrumentation for an inter-strand resistance measurement in a Rutherford cable with 36 strands. Image by Devred et al. [110]. (b) The same method applied to a Roebel cable.

The network of inter-strand resistance measurements in cables with four and six strands are shown in figure 3.13. A current I is applied between strands N and $\frac{1}{2}N$. Assuming a cable length ℓ, the average transverse current per unit-length is I/ℓ. Note that this representation is only valid for currents below the critical current, because the potential difference within the strands is assumed to be zero.

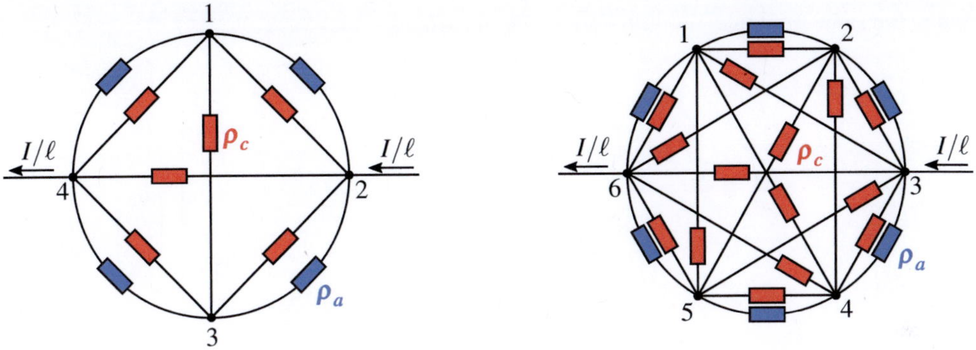

Figure 3.13: Electrical network representation of inter-strand resistance measurements in cables with four and six strands. The nodes represent the superconducting strands, which have no internal potential difference. Each pair of strands is connected by ρ_c, while neighbouring strands are connected by ρ_a as well. A current density I/ℓ is applied between strands N and $\frac{1}{2}N$.

The potentials $U_1, U_2, \ldots, U_N$ in each node can be found using Kirchhoff's current law, which states that the sum of currents flowing into a node equals the sum of currents flowing out of that node. For a four-strand cable we find:

$$\text{Node 1:} \quad \frac{U_2-U_1}{\rho_a}+\frac{U_4-U_1}{\rho_a}+\frac{U_2-U_1}{\rho_c}+\frac{U_3-U_1}{\rho_c}+\frac{U_4-U_1}{\rho_c}=0 \tag{3.10}$$

$$\text{Node 2:} \quad \frac{U_1-U_2}{\rho_a}+\frac{U_3-U_2}{\rho_a}+\frac{U_1-U_2}{\rho_c}+\frac{U_3-U_2}{\rho_c}+\frac{U_4-U_2}{\rho_c}+\frac{I}{\ell}=0 \tag{3.11}$$

$$\text{Node 3:} \quad \frac{U_2-U_3}{\rho_a}+\frac{U_4-U_3}{\rho_a}+\frac{U_1-U_3}{\rho_c}+\frac{U_2-U_3}{\rho_c}+\frac{U_4-U_3}{\rho_c}=0 \tag{3.12}$$

Taking $U_4=0$ as a reference this can be written as:

$$-\left(\frac{2}{\rho_a}+\frac{3}{\rho_c}\right)U_1+\left(\frac{1}{\rho_a}+\frac{1}{\rho_c}\right)U_2+\frac{1}{\rho_c}U_3=0 \tag{3.13}$$

$$\left(\frac{1}{\rho_a}+\frac{1}{\rho_c}\right)U_1-\left(\frac{2}{\rho_a}+\frac{3}{\rho_c}\right)U_2+\left(\frac{1}{\rho_a}+\frac{1}{\rho_c}\right)U_3=-I/\ell \tag{3.14}$$

$$\frac{1}{\rho_c}U_1+\left(\frac{1}{\rho_a}+\frac{1}{\rho_c}\right)U_2-\left(\frac{2}{\rho_a}+\frac{3}{\rho_c}\right)U_3=0 \tag{3.15}$$

In matrix form:

$$\begin{bmatrix} -\left(\frac{2}{\rho_a}+\frac{3}{\rho_c}\right) & \left(\frac{1}{\rho_a}+\frac{1}{\rho_c}\right) & \frac{1}{\rho_c} \\ \left(\frac{1}{\rho_a}+\frac{1}{\rho_c}\right) & -\left(\frac{2}{\rho_a}+\frac{3}{\rho_c}\right) & \left(\frac{1}{\rho_a}+\frac{1}{\rho_c}\right) \\ \frac{1}{\rho_c} & \left(\frac{1}{\rho_a}+\frac{1}{\rho_c}\right) & -\left(\frac{2}{\rho_a}+\frac{3}{\rho_c}\right) \end{bmatrix} \begin{bmatrix} U_1 \\ U_2 \\ U_3 \end{bmatrix} = \begin{bmatrix} 0 \\ -I/\ell \\ 0 \end{bmatrix} \tag{3.16}$$

Using the same principle, one can find for the six-strand cable:

$$\begin{bmatrix} -\left(\frac{2}{\rho_a}+\frac{5}{\rho_c}\right) & \left(\frac{1}{\rho_a}+\frac{1}{\rho_c}\right) & \frac{1}{\rho_c} & \frac{1}{\rho_c} & \frac{1}{\rho_c} \\ \left(\frac{1}{\rho_a}+\frac{1}{\rho_c}\right) & -\left(\frac{2}{\rho_a}+\frac{5}{\rho_c}\right) & \left(\frac{1}{\rho_a}+\frac{1}{\rho_c}\right) & \frac{1}{\rho_c} & \frac{1}{\rho_c} \\ \frac{1}{\rho_c} & \left(\frac{1}{\rho_a}+\frac{1}{\rho_c}\right) & -\left(\frac{2}{\rho_a}+\frac{5}{\rho_c}\right) & \left(\frac{1}{\rho_a}+\frac{1}{\rho_c}\right) & \frac{1}{\rho_c} \\ \frac{1}{\rho_c} & \frac{1}{\rho_c} & \left(\frac{1}{\rho_a}+\frac{1}{\rho_c}\right) & -\left(\frac{2}{\rho_a}+\frac{5}{\rho_c}\right) & \left(\frac{1}{\rho_a}+\frac{1}{\rho_c}\right) \\ \frac{1}{\rho_c} & \frac{1}{\rho_c} & \frac{1}{\rho_c} & \left(\frac{1}{\rho_a}+\frac{1}{\rho_c}\right) & -\left(\frac{2}{\rho_a}+\frac{5}{\rho_c}\right) \end{bmatrix} \begin{bmatrix} U_1 \\ U_2 \\ U_3 \\ U_4 \\ U_5 \end{bmatrix} = \begin{bmatrix} 0 \\ 0 \\ -I/\ell \\ 0 \\ 0 \end{bmatrix} \tag{3.17}$$

In general, for any even number of strands:

$$\sigma \mathbf{U} = \mathbf{J} \tag{3.18}$$

$$\text{where} \quad \sigma_{ij} = \begin{cases} -\left(\frac{2}{\rho_a}+\frac{N-1}{\rho_c}\right) & i=j \\ \frac{1}{\rho_a}+\frac{1}{\rho_c} & |i-j|=1 \\ \frac{1}{\rho_c} & |i-j|>1 \end{cases} \tag{3.19}$$

$$J_i = \begin{cases} -I/\ell & i=\frac{1}{2}N \\ 0 & i \neq \frac{1}{2}N \end{cases} \tag{3.20}$$

This linear set of equations can be solved in a program such as Matlab to find the voltage profile for any ρ_a and ρ_c.

Figure 3.14 shows the calculated voltage profiles for 10-strand cables with different ratios of ρ_c/ρ_a. The shape of the profile depends on this ratio. If $\rho_c \ll \rho_a$, the current flows directly from strand 5 to strand 10. Between other pairs of strand, the current is negligible, and therefore they have the same electric potential. The resulting voltage profile has two spikes at strand 5 and 10, while all other strands are at an intermediate level. If on the other hand $\rho_c \gg \rho_a$, current transfer can only occur between adjacent strands. The applied current will pass through all strands, leading to a voltage profile shaped like a triangle. Intermediate ratios of ρ_c/ρ_a result in voltage profiles with features of both extreme cases.

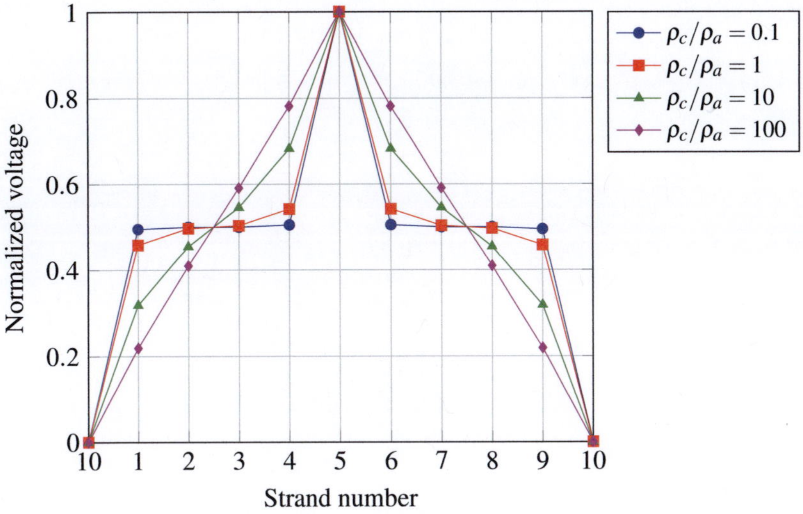

Figure 3.14: Calculated voltage profiles for inter-strand resistance measurements in 10-strand cables with different ratios of ρ_c/ρ_a. The current is applied from strand 5 to strand 10. The voltage profile is normalized to the voltage in strand 5.

3.5.2 Sample preparation

All tested samples were prepared from 12 mm wide coated conductors. The conductors were punched with a transposition length of 126 mm and a strand width of 5.5 mm. After punching, the strands were cleaned with acetone. Each cable was assembled from 10 strands and had a length of about two transpositions. A pressure is applied to one transposition (126 mm) of the cable as seen in figure 3.15. Outside of the pressed area, the strands are electrically insulated from each other by means of a polyimide tape. This ensures that current transfer between strands occurs only within the pressed part. At the insulated cable end, current leads and voltage taps are connected. By injecting the current away from the pressed area, it is ensured that all current enters the superconductor.

In addition to these pressed cables, two soldered cables were prepared. The strands of these cables were pre-soldered with In52Sn48 using the procedure described in section 3.4.1. After cable assembly, the strands were soldered together over one transposition length. This was done using the aluminium soldering tool (figure 3.7) under a force of 940 N.

In every measurement a current of 10 A is applied to strand 5 and returned through strand 10. The voltage on each strand is measured relative to strand 10 using a nanovoltmeter and a multiplexer.

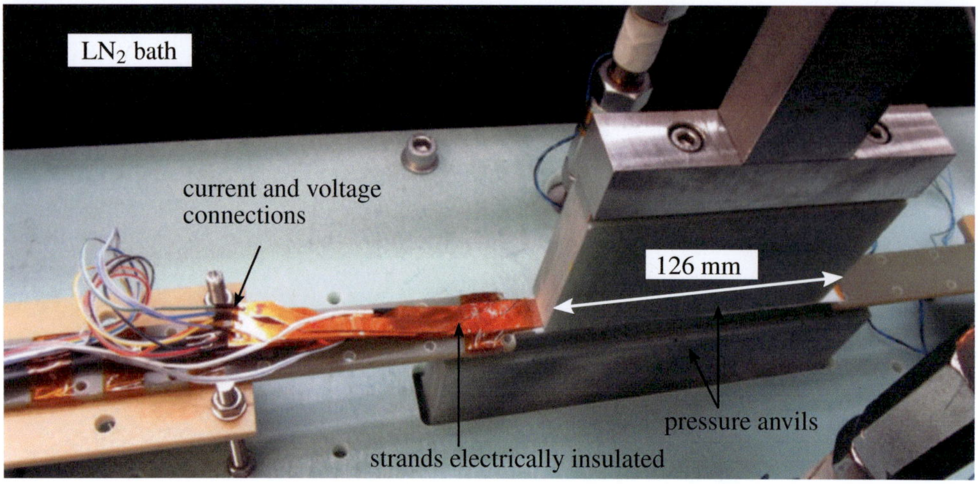

Figure 3.15: A Roebel cable prepared for an inter-strand resistance measurement. Using the hydraulic press, a pressure is applied to one transposition length (126 mm). The strands are electrically insulated outside of the pressed area.

3.5.3 Results: Roebel cables with SuperPower conductor

Four cables were prepared from the same SuperPower tape as used for the two-tape contacts in section 3.4.2. Two of the four cables were soldered. The measured voltage profiles at different levels of applied pressure are shown in figures 3.16 and 3.17. The force is normalized to the entire pressed area of the cable of 126 mm × 12 mm. Because Roebel cables have an uneven thickness, the computed stress represents only an average value. As in the soldered two-tape contacts, the voltage profile of the soldered cables does not depend on pressure. The shape of the voltage profile indicates an intermediate ratio ρ_c/ρ_a. In the pressed cables without solder, the voltage decreases with increasing pressure. The shape voltage profile is more irregular than in the soldered cables. This indicates a larger spread of inter-strand resistances within the cable sample.

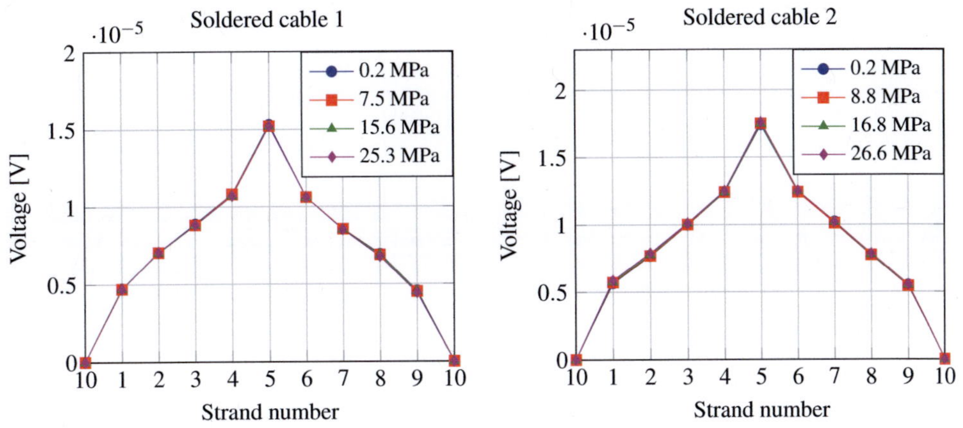

Figure 3.16: Voltage profiles measured in the two soldered cables at different levels of applied pressure. The lines connecting measurements points are only a guidance to the eyes.

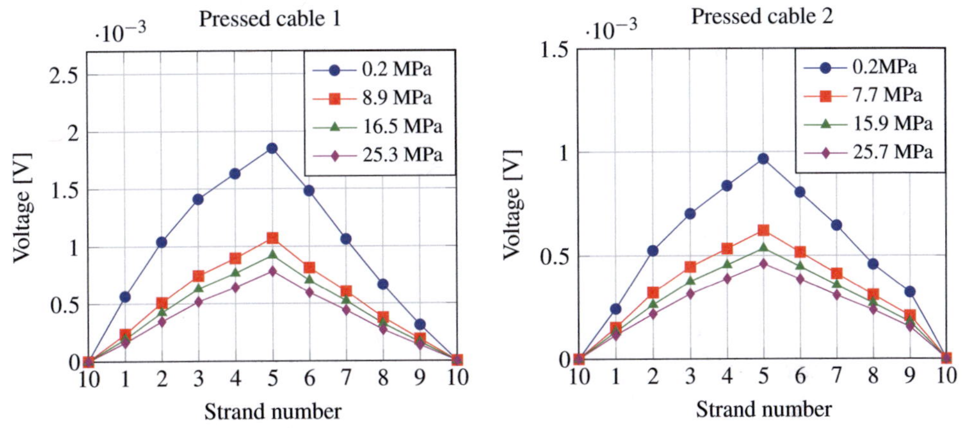

Figure 3.17: Voltage profiles measured in the two pressed cables at different levels of applied pressure. The lines connecting measurements points are only a guidance to the eyes.

The values of ρ_a and ρ_c are determined by fitting the measurement data to the theoretical profile described by equation 3.18. This is done using Matlab's non-linear least-squares fitting function "lsqnonlin". Figure 3.18 shows the results of this fitting procedure for selected data. For the soldered cable, a very good match to the theoretical profile is found with $\rho_a = 0.193$ μΩm and $\rho_c = 2.45$ μΩm. In the pressed cable, $\rho_a = 4.67$ μΩm and $\rho_c \gg \rho_a$ give a reasonable match. The value of ρ_c being much larger than ρ_a is typical for cables in which strands are connected to their neighbours only. The good match supports the validity of the Rutherford inter-strand resistance model for Roebel cables.

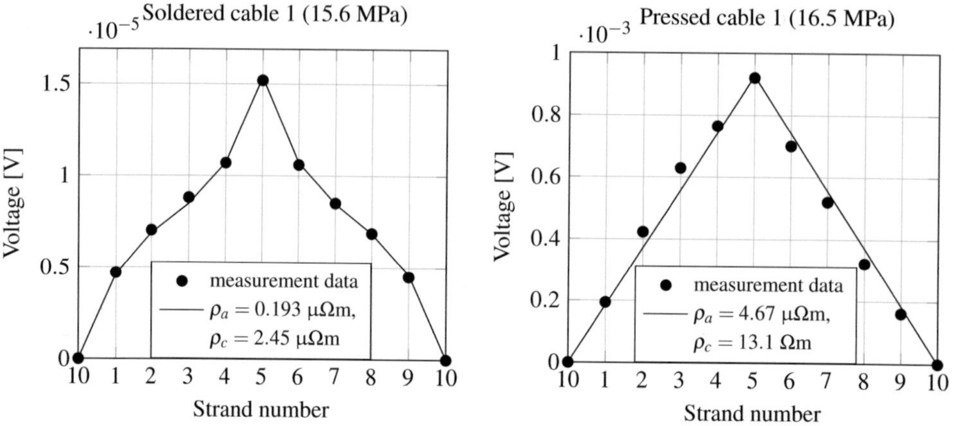

Figure 3.18: Selected measurement data with corresponding least-squares fit of the Rutherford inter-strand resistance model.

The values of ρ_a and ρ_c as a function of average transverse stress are shown in figure 3.19. For the pressed cables, only ρ_a values are shown. Because ρ_c is much larger than ρ_a, small changes in ρ_c hardly influence the voltage profile, and a precise value cannot be determined. Besides, the precise value of ρ_c is not relevant if ρ_c much larger than ρ_a, because current transfer through ρ_c will be negligible. The pressed cables show a clear pressure dependency of ρ_a while soldered cables have pressure independent inter-strand resistance. This reflects the behaviour of pressed and soldered contacts. There is a noticeable difference between samples prepared with the same method. Pressed cable 2 has a ρ_a value approximately twice that of pressed cable 1. The ρ_a and ρ_c values of the soldered cables differ up to 20% between samples.

A comparison to the two-tape contacts is made by normalizing ρ_a to the contact area. For this, the adjacent strand contact resistance for an entire transposition length ρ_a/ℓ_t is multiplied with the theoretical contact area of 552 mm^2 (see table 3.1 for the area calculation). The resulting contact surface resistivities are compared to simple front-to-back contacts in figure 3.20. For the soldered cables, a similar order of resistivity is found as in the two-tape contacts. The pressed cables however have higher resistivity than pressed contacts. It is important to consider that a transverse force on a Roebel cable is not homogeneously distributed. Parts of the contact areas experiencing high transverse stress likely have a lower resistance than parts with low stress levels. As shown by Fleiter et al. [111], the effective section of a Roebel cable with an even number of strands is:

$$E_{s,even} = \frac{2N}{\ell_t W \tan(\alpha)} \left(\frac{w_c \sin(\alpha)}{\sin(2\alpha)} - \frac{W}{2} + w_s \right)^2 \tag{3.21}$$

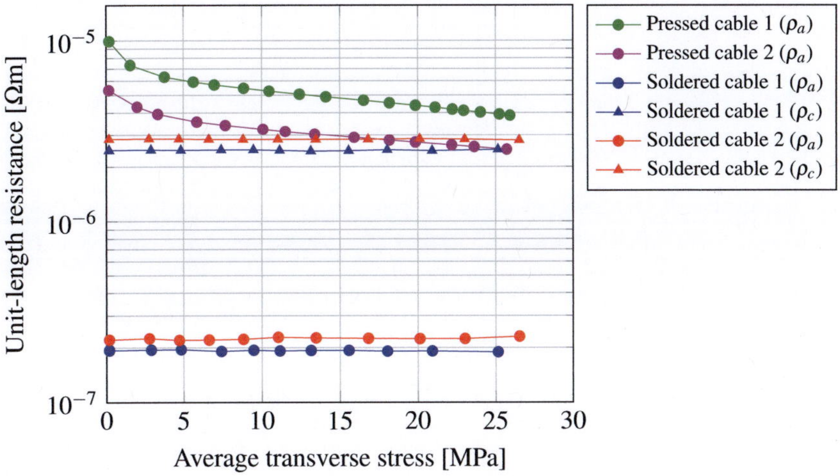

Figure 3.19: Measured ρ_a and ρ_c values of pressed and soldered Roebel cables as a function transverse stress.

The effective section is defined as the fraction of the cable area where the cable thickness equals the maximum thickness. For the cable architecture used here, Fleiter's formula predicts that a transverse pressure is concentrated on 16.4% percent of the cable surface area. The overlap area of two adjacent strand is located in only one half of the cable, and it therefore experiences a stress on an area 8.2% of the cable surface area. This contact area between adjacent strands experiencing pressure is:

$$0.082 \times 126 \text{ mm} \times 12 \text{ mm} \approx 124 \text{ mm}^2 \tag{3.22}$$

which is significantly smaller than the total overlap area of 552 mm^2. When both force and resistance are normalized taking into account the effective section, the resistivities show better agreement with two-tape contacts (see figure 3.20). However, the sample-to-sample variation is so large that no definite conclusion can be made. For prediction of inter-strand resistance in pressed cables it is therefore not recommended to rely on two-tape press contacts only.

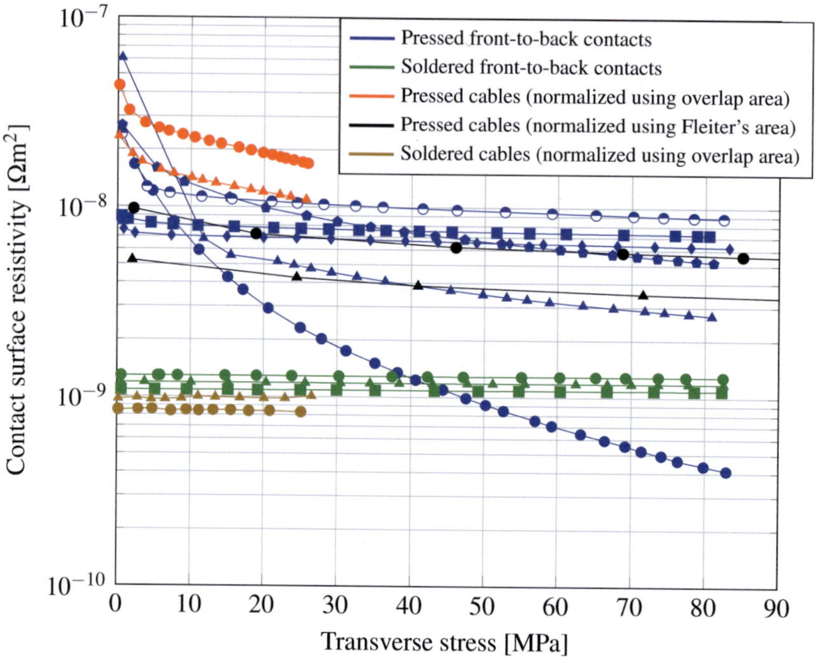

Figure 3.20: Contact surface resistivity as function of transverse stress for Roebel cables and two-tape contacts made from SuperPower coated conductors. Graphs with the same color represent different samples prepared in the same way.

After the inter-strand resistance measurements, the cables were embedded in epoxy and polished cross-sections were made. Microscopic images of a pressed cable and a soldered cable are shown in figures 3.21 and 3.22. The thickness of the tapes is found to vary over the width. It is 76.4 μm near the edges while the average thickness is 94.9 μm closer to the middle. Such a difference is enough to cause gaps between the strands, reducing the contact area. In the soldered cable, one can see that the space between the strands is mostly filled with solder, although some gaps are still visible. The "central hole" of the cable is also filled with solder, connecting strands on opposite sides of the cable. This may explain why the soldered cables had ρ_a and ρ_c of similar order, while in pressed cables ρ_c is much larger than ρ_a.

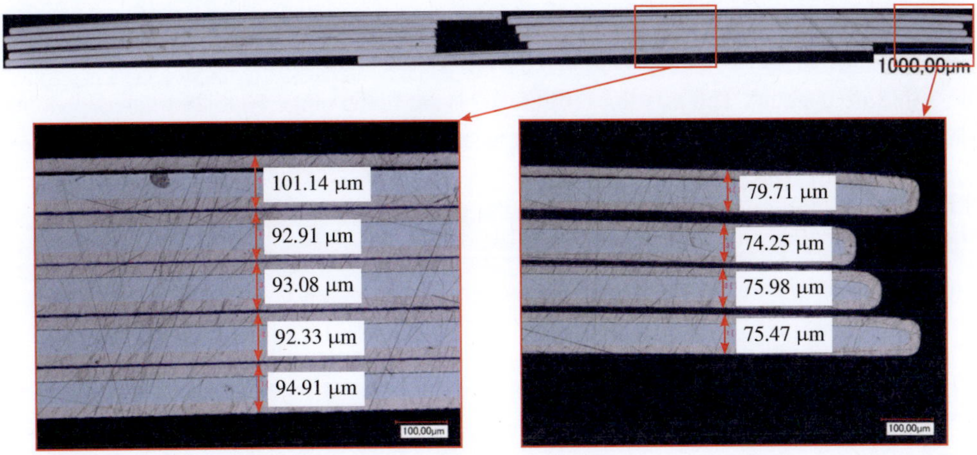

Figure 3.21: Cross-section of pressed cable 1. The tapes have an average thickness of 76.4 µm near the edges while the average thickness is 94.9 µm closer to the middle.

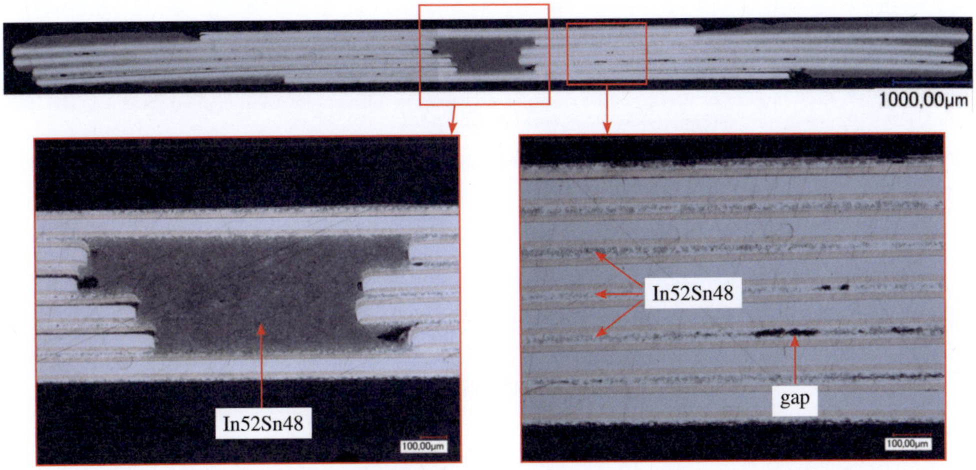

Figure 3.22: Cross-section of soldered cable 1. The central hole in of the cable is filled with In52Sn48 solder at the site of this cross-section

3.5.4 Results: Roebel cables with SuperOx conductor

Two Roebel cables without solder were prepared from SuperOx coated conductor. It was planned to prepare several soldered cables as well. However, attempts to prepare Roebel cables strands from solder-coated tape were not successful because of a delamination problem.

Voltage profiles at different pressure levels for both cables are shown in figure 3.23. The irregular shape of the profile indicates that the resistances of the different press contacts within the cable are unequal. The ρ_a values (figure 3.24) are in the range 10 to 28 µΩm depending on pressure, somewhat higher than those observed in cables from SuperPower conductor (2.5 – 10 µΩm). The difference between the two samples is remarkably small.

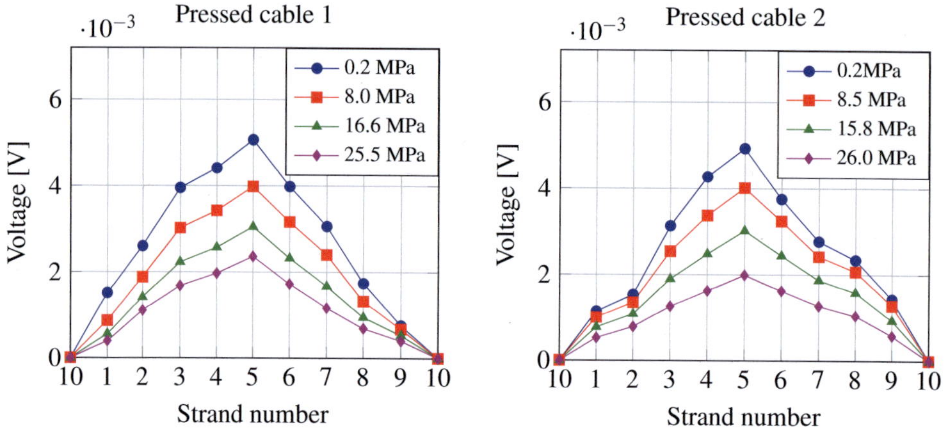

Figure 3.23: Voltage profiles measured in the two pressed cables at different levels of applied pressure. The lines connecting measurements points are only a guidance to the eyes.

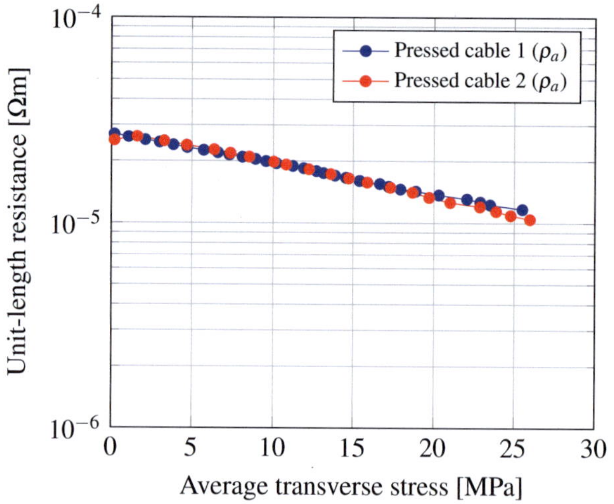

Figure 3.24: Measured ρ_a values of pressed Roebel cables made from SuperOx tapes as a function transverse stress.

3.6 Conclusion

The resistance of two-tape contacts and inter-strand resistance in Roebel cables were investigated at 77 K and as a function of pressure. Soldered two-tape contacts of SuperPower tapes were found to have a surface resistivity of in the range $1.3 - 2.2 \cdot 10^{-11}$ Ωm^2 in front-to-front orientation and $1.0 - 1.3 \cdot 10^{-9}$ Ωm^2 in front-to-back orientation. This difference of two orders of magnitude can be ascribed to the internal front-to-back resistance. Contacts without solder had a pressure-dependent resistance and a large spread between samples.

The inter-strand resistance was investigated using an established method for Rutherford cables. This method involves measuring the voltage profile under an applied transverse current, followed by fitting to an inter-strand resistance model of two parameters (R_a and R_c). The inter-strand resistance model was slightly modified to better suit Roebel cables. The modified model parameters are ρ_a, the unit-length resistance connecting adjacent strands, and ρ_c, the unit-length connecting each pair of strands. A good fit could be reached by fitting calculated voltage profiles to measurement data using ρ_a and ρ_c as fitting parameters. This supports the use of the Rutherford methods for Roebel cables. The inter-strand resistance was found to be $\rho_a = 0.19 - 0.23$ μΩm, $\rho_c = 2.4 - 2.9$ μΩm for solder-filled cables from SuperPower conductor. For pressed cables, $\rho_a = 2.5 - 11$ μΩm depending on pressure, and ρ_c is much larger than ρ_a. Slightly higher values of $\rho_a = 10 - 28$ μΩm were found in cables from SuperOx conductor. All measured ρ_a and ρ_c values are not universal but depend on the cable parameters.

When ρ_a values of soldered cables are normalized using the overlap area of adjacent strands, a reasonable match with front-to-back soldered contacts is found. For pressed cables, such a normalization is not easily done, because the overlap area does not experience a homogeneous pressure. Also, variations in tape thickness can lead to variations in contact area which cannot be controlled.

Chapter 4

AC loss in partially coupled cables

4.1 Introduction

Although superconductors have zero resistivity, dissipation can still arise if time-dependent currents or magnetic fields are involved. In type-II superconductors such as *RE*BCO, the magnetic field penetrates in the form of Abrikosov current vortices. A change in magnetic field leads to a change in vortex density. The vortex lattice moves in a viscous way, and thus work is done on the superconductor. The resulting dissipation is known as AC loss. In the macroscopic view, AC loss can also be viewed as dissipation that arises as the current is driven over the critical value by an induced electric field:

$$\nabla \times \mathbf{E} = -\frac{\partial \mathbf{B}}{\partial t} \tag{4.1}$$

Both views are in fact equivalent, and calculations using either model give the same results [109, p. 159].

In superconducting devices, AC losses are often the main contribution to the heat load. Since the efficiency of low-temperature cooling is limited by Carnot's theorem, the increase in power consumption is larger than the AC loss itself. Additionally, the AC loss can increase temperature and cause stability problems. Consideration of AC loss is therefore essential for the design of an efficient and stable superconducting device.

AC loss originates from different mechanisms called hysteresis loss and coupling loss. Hysteresis loss relates to induced currents that flow entirely in superconducting material. In a superconductor described by the critical state model, these currents do not decay. The hysteresis loss per cycle is therefore independent of the time taken to complete the cycle. In a wire or cable with multiple superconducting filaments, coupling currents that flow between filaments can be induced. These currents flow partially in superconducting and resistive material. Because coupling currents experience a resistance, they decay with time. Coupling losses are therefore frequency dependent. In a cable, the inter-strand resistance and transposition length influence the magnitude of coupling currents and loss. A third contribution to losses can result from eddy currents flowing entirely in resistive material.

AC loss in Roebel cables has been investigated in several studies. A comprehensive overview can be found in a review paper by Goldacker et al. [22]. However, only a few publications pay special attention to inter-strand resistance and its effect on coupling loss. In an early work by Schuller et al., a partially coupled Roebel cable was prepared by impregnation with a conductive silver epoxy [112]. An inter-strand resistance value of 3.6 $\mu\Omega \pm 40\%$ was found by measuring the resistance between neighbouring strands and taking the average. The magnetization AC losses of the sample were measured with frequencies up to 800 Hz. For frequencies below 200 Hz, a linear increase with frequency was observed and attributed to coupling loss. At higher frequencies, the AC loss was below the linear trend. The authors claim this results from decreased hysteresis loss due to shielding by coupling currents. In a follow-up investigation, AC losses of Roebel cables impregnated with a silver-filled and unfilled epoxy resin were compared [113]. The sample impregnated with silver-filled resin was found to a have slightly higher frequency dependent AC loss. In both works no quantitative analysis of the relation between inter-strand resistance and AC loss was made.

Lakshmi et al. introduced a connection between neighbouring strands by soldering copper "bridges" on top of the cable [114]. The resistance of the copper bridges was estimated to be in the range 5-15 $\mu\Omega$ from electrical measurements on single bridges. The coupling loss was determined by measuring the magnetization loss as a function of frequency and subtracting an extrapolated value at zero frequency. The resistance of the copper bridges was then determined from coupling loss by fitting the data to a coupling loss equation proposed earlier by the same author [115]. A value of 21 $\mu\Omega$ was found, only slightly higher than the electrically measured value. This work was the first to demonstrate a quantitative relation between inter-strand resistance and coupling loss in Roebel cables.

In this chapter, the aim is to systematically investigate the relation between inter-strand resistance and AC loss. The focus lies on magnetization losses in a homogeneous magnetic field that changes sinusoidally in time and is oriented perpendicularly to the wide face of the cable. The AC loss taking into account inter-strand coupling is calculated using the inter-strand resistance model described in the previous chapter. In section 4.2, the coupling loss is estimated using a simplified representation of two parallel bars. In section 4.3, a numerical method is proposed to calculate the AC loss in a cable with variable inter-strand resistance. To verify the validity of the calculations, the AC loss is measured in cables with different inter-strand resistance using the calibration-free method. These measurements are described in section 4.4. The measurement data are compared to both calculation methods.

4.2 Calculation of coupling loss using a simplified cable geometry

This section aims to find an estimation of the coupling loss in Roebel cables without numerical techniques. To do this, the geometry of the cable is strongly simplified as shown in figure 4.1. The cable is represented by two parallel rectangular bars of infinite length. A time-dependent external magnetic field $B_{ext}(t)$ is applied perpendicular to the wide face. A coupling current $I(t)$ is induced, and flows in opposite directions in the two bars. This current is assumed to be evenly distributed over the bar cross-section. Since the strands are transposed, the coupling current transfers from strand-to-strand and experiences a resistance due to inter-strand contacts. This effect is introduced by giving the bars an effective resistance per unit length of r_e with units of Ω/m. An equation relating r_e and the inter-strand resistance will be derived in 4.2.1.

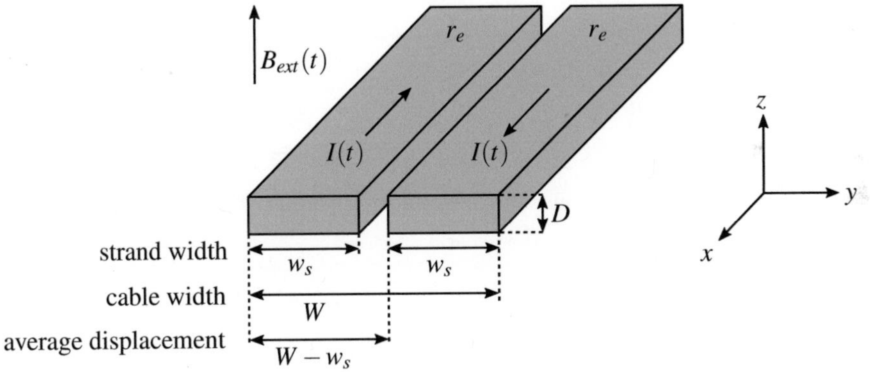

Figure 4.1: Simplified geometry for estimation of the coupling loss in a perpendicular field. The two halves of the Roebel cables are replaced by two rectangular bars with an effective resistance r_e per unit length. The current density in the bar is assumed to be homogeneous. The current of the two bars is displaced by $W - w_s$ in the y-direction.

This representation does not take into account the superconducting properties of the cable. Interaction between coupling currents and superconductor magnetization are therefore not considered. This effect will be included in the more elaborate numerical approach described in section 4.3.

The bars are displaced by $W - w_s$ in the y-direction. Considering a rectangular loop of this width extending along the cable, a voltage per unit length

$$(W - w_s)\frac{\mathrm{d}B_{ext}}{\mathrm{d}t} \tag{4.2}$$

is induced according Faraday's law. The current $I(t)$ experiences a resistive voltage of $r_e I(t)$ in both cable halves and an inductive voltage $L\mathrm{d}I/\mathrm{d}t$, where L is the self-inductance of the loop per unit length. Equating the voltage per unit length, one finds the following differential equation for the current:

$$(W - w_s)\frac{\mathrm{d}B_{ext}}{\mathrm{d}t} = 2r_e I(t) + L\frac{\mathrm{d}I}{\mathrm{d}t} \tag{4.3}$$

If the magnetic field B_{ext} is constant in time, this equations has solutions in the form

$$I(t) = I_0 e^{-t/\tau} \tag{4.4}$$

where τ is a time constant given by

$$\tau = \frac{L}{2r_e} \tag{4.5}$$

To find $I(t)$ in a sinusoidal magnetic field with amplitude B_0 and angular frequency $\omega = 2\pi f$, substitute the following expression for the magnetic field:

$$B_{ext}(t) = B_0 \sin(\omega t) = \frac{B_0}{2i}\left(e^{i\omega t} - e^{-i\omega t}\right) \tag{4.6}$$

$$\frac{\mathrm{d}B_{ext}}{\mathrm{d}t} = B_0\omega\cos(\omega t) = \frac{B_0\omega}{2}\left(e^{i\omega t} + e^{-i\omega t}\right) \tag{4.7}$$

At low frequencies, where self-inductance does not play a role, the current is expected to be in phase with the induced voltage. For the general case, one looks for a solution with a phase shift δ relative to $\mathrm{d}B_{ext}/\mathrm{d}t$:

$$I(t) = I_0\cos(\omega t - \delta) = \frac{I_0}{2}\left(e^{i(\omega t-\delta)} + e^{-i(\omega t-\delta)}\right) \tag{4.8}$$

$$\frac{\mathrm{d}I}{\mathrm{d}t} = -I_0\omega\sin(\omega t - \delta) = -\frac{I_0\omega}{2i}\left(e^{i(\omega t-\delta)} - e^{-i(\omega t-\delta)}\right) \tag{4.9}$$

Substitution of equations 4.7-4.9 into the differential equation 4.5 gives:

$$(W - w_s)\frac{B_0\omega}{2}\left(e^{i\omega t} + e^{-i\omega t}\right) = r_e I_0\left(e^{i(\omega t-\delta)} + e^{-i(\omega t-\delta)}\right) - \frac{I_0\omega L}{2i}\left(e^{i(\omega t-\delta)} - e^{-i(\omega t-\delta)}\right) \tag{4.10}$$

$$= e^{i\omega t}\left(r_e I_0 e^{-i\delta} - \frac{I_0\omega L}{2i} e^{-i\delta}\right) + e^{-i\omega t}\left(r_e I_0 e^{i\delta} + \frac{I_0\omega L}{2i} e^{i\delta}\right) \tag{4.11}$$

$$\Rightarrow \begin{cases} (W-w_s)\dfrac{B_0\omega}{2} = \left(r_e I_0 - \dfrac{I_0\omega L}{2i}\right) e^{-i\delta} & (4.12) \\ (W-w_s)\dfrac{B_0\omega}{2} = \left(r_e I_0 + \dfrac{I_0\omega L}{2i}\right) e^{i\delta} & (4.13) \end{cases}$$

The phase shift δ can be eliminated by multiplying equations 4.12 and 4.13:

$$(W-w_s)^2\left(\frac{B_0\omega}{2}\right)^2 = I_0^2\left(r_e - \frac{\omega L}{2i}\right)\left(r_e + \frac{\omega L}{2i}\right) = I_0^2\left(r_e^2 + \left(\frac{\omega L}{2}\right)^2\right) \tag{4.14}$$

The amplitude of the coupling current is found by solving this equation for I_0:

$$I_0 = \frac{(W-w_s)B_0\omega}{2\sqrt{r_e^2 + \left(\frac{\omega L}{2}\right)^2}} \tag{4.15}$$

The time averaged power per unit length is:

$$P = 2I_{RMS}^2 r_e = I_0^2 r_e = \frac{(W-w_s)^2 B_0^2\omega^2 r_e}{4\left(r_e^2 + \left(\frac{\omega L}{2}\right)^2\right)} = \frac{(W-w_s)^2 B_0^2\omega^2}{4r_e(1+(\omega\tau)^2)} = \frac{(W-w_s)^2 B_0^2\pi^2 f^2}{r_e(1+(2\pi f\tau)^2)} \tag{4.16}$$

where the factor 2 is due to heating in both cable halves. The coupling loss per cycle Q is found by dividing the power by the frequency:

$$Q = \frac{P}{f} = \frac{(W-w_s)^2 B_0^2\pi^2 f}{r_e(1+(2\pi f\tau)^2)} \tag{4.17}$$

The phase shift δ can be found by subtracting equations 4.12 and 4.13 and solving for δ:

$$\left(r_e - \frac{\omega L}{2i}\right) e^{-i\delta} - \left(r_e + \frac{\omega L}{2i}\right) e^{i\delta} = 0 \tag{4.18}$$

$$-r_e\left(e^{i\delta} - e^{-i\delta}\right) - \frac{\omega L}{2i}\left(e^{i\delta} + e^{-i\delta}\right) = 0 \tag{4.19}$$

$$-2r_e\frac{e^{i\delta} - e^{-i\delta}}{2i} + \omega L\frac{e^{i\delta} + e^{-i\delta}}{2} = 0 \tag{4.20}$$

$$-2r_e\sin(\delta) + \omega L\cos(\delta) = 0 \tag{4.21}$$

$$\frac{\sin(\delta)}{\cos(\delta)} = \tan(\delta) = \frac{\omega L}{2r_e} \tag{4.22}$$

$$\Rightarrow \delta = \tan^{-1}\left(\frac{\omega L}{2r_e}\right) = \tan^{-1}(\omega\tau) \tag{4.23}$$

In the low frequency limit for which f is much smaller than $(2\pi\tau)^{-1}$, the loss per cycle is proportional to the frequency:

$$Q_{low} = \frac{(W - w_s)^2 B_0^2 \pi^2 f}{r_e} \tag{4.24}$$

This equation was also derived by Lakshmi et al. [115] in a different form. The authors also noted the similarity to the equation for coupling losses of Rutherford cables in a *parallel* transverse field [107], which underlines the idea that Roebel and Rutherford are topologically identical forms of a single layer transposed cable. The phase shift in the low frequency limit is 0°, and the value of time constant does not influence the loss level. In the high frequency limit where f is much larger than $(2\pi\tau)^{-1}$, the loss per cycle is inversely proportional to the frequency:

$$Q_{high} = \frac{(W - w_s)^2 B_0^2}{4 r_e f \tau^2} = \frac{(W - w_s)^2 B_0^2 r_e}{f L^2} \tag{4.25}$$

and the phase shift is 90°. Figure 4.2 shows the frequency dependence of loss per cycle Q, dissipative power P and phase lag δ over a wide range of frequencies. Although the loss per cycle has a maximum at a frequency of $2\pi\tau$, the power increases further and becomes constant in the high frequency limit.

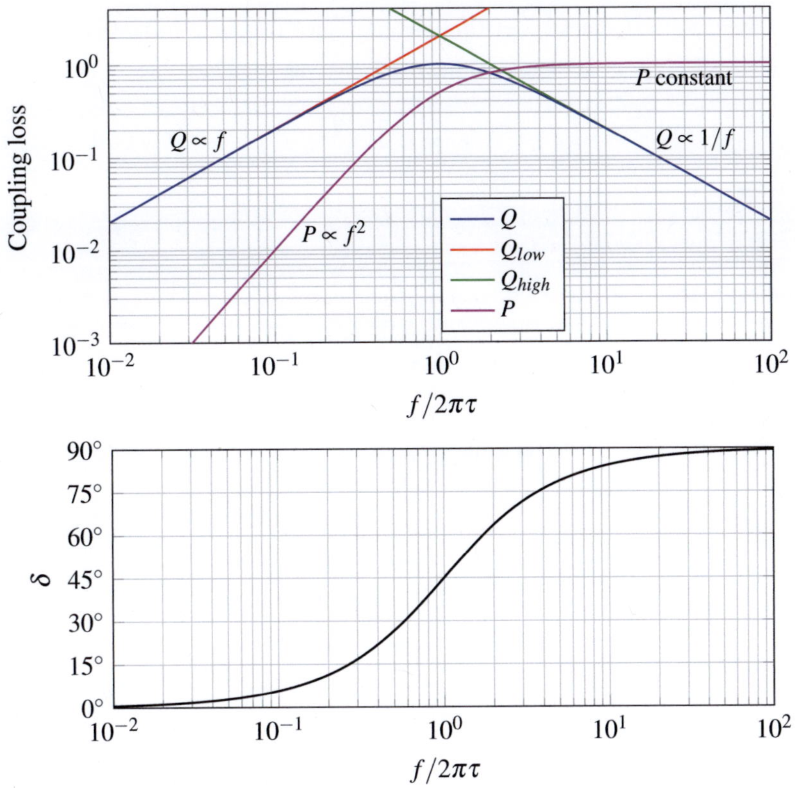

Figure 4.2: Frequency dependence of the coupling loss per cycle Q [J/m], the power P [W/m] and the phase shift δ calculated using equations 4.16, 4.17 and 4.23. The coupling loss and power are normalized to the peak values. For these calculations, the Roebel cable was approximated by two parallel rectangular bars of infinite length.

4.2.1 Effective resistance and self-inductance

The values of the effective resistance r_e and self-inductance L per unit length will be derived in this section. These values are needed to evaluate equation 4.17 for the coupling loss per cycle.

The strands of a Roebel cable cross to the other cable side each half transposition length. Therefore, the coupling current transfers from strand to strand through physical contacts of half a transposition length. This is schematically shown in figure 4.3. For each transfer to the next strand, the current experiences a resistance of

$$\frac{2\rho_a}{\ell_t} \tag{4.26}$$

where ρ_a is the resistance between adjacent strands per unit length (see also section 3.5). The current transfers N times per transposition length ℓ_t, where N is the number of strands.

Therefore, there are N/ℓ_t transfers per unit length. The effective resistance is the product of the number of transfers per unit length and the resistance per transfer:

$$r_e = \frac{2\rho_a N}{\ell_t^2} \tag{4.27}$$

The effective resistance r_e is a resistance per unit length, and thus has units of Ω/m.

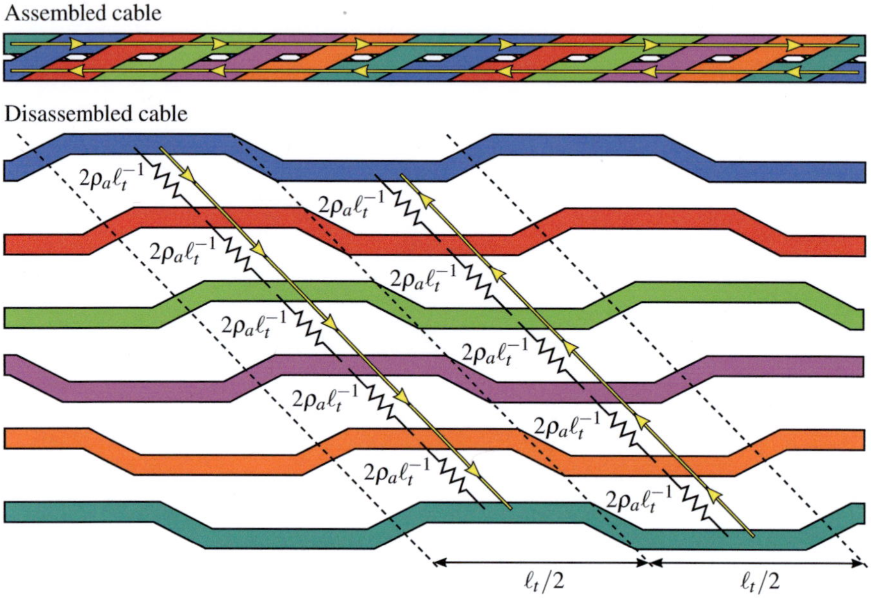

Figure 4.3: Top view of a Roebel cable with coupling currents schematically drawn by yellow arrows. The cable is disassembled to show how the current transfers from strand to strand over half a transposition length.

The self and mutual inductances of linear conductors were calculated by E. Rosa [116]. According this work, the self-inductance L_1 of a rectangular bar can be approximated by:

$$L_1 \approx 2\mu_0\ell\left(\ln\left(\frac{2\ell}{\alpha+\beta}\right)+\frac{1}{2}+\frac{0.2235(\alpha+\beta)}{\ell}\right) \tag{4.28}$$

and the mutual inductance M of two parallel conductors by:

$$M \approx 2\mu_0\ell\left(\ln\left(\frac{2\ell}{d}\right)-1+\frac{d}{l}\right) \tag{4.29}$$

In these equations, α and β are the lengths of the sides of rectangular bar cross-section, ℓ is the length of the bar, and d is the distance between the two parallel conductors. The self-inductance of a return circuit of two parallel bars is:

$$2L_1 + 2M = 4\mu_0\ell\left(\ln\left(\frac{2l}{\alpha+\beta}\right) + \frac{1}{2} + \frac{0.2235(\alpha+\beta)}{\ell} - \ln\left(\frac{2\ell}{d}\right) + 1 - \frac{d}{\ell}\right) \tag{4.30}$$

$$= 4\mu_0\ell\left(\ln\left(\frac{d}{\alpha+\beta}\right) + \frac{3}{2} + \frac{0.2235(\alpha+\beta)-d}{\ell}\right) \tag{4.31}$$

For a long conductor in which ℓ is much larger than α, β, and d, the last term becomes zero and the inductance per unit length can be simplified to:

$$\frac{2L_1 + 2M}{\ell} \approx \mu_0\left(4\ln\left(\frac{d}{\alpha+\beta}\right) + 6\right) \tag{4.32}$$

To find the self-inductance per unit length for the simplified geometry in figure 4.1 one substitutes $\alpha = w_s$, $\beta = D$ and $d = W - w_s$:

$$L \approx \mu_0\left(4\ln\left(\frac{W - w_s}{w_s + D}\right) + 6\right) \tag{4.33}$$

The effective resistance and self-inductance are used to evaluate equation 4.17 for the coupling loss per cycle.

4.2.2 Effect of finite sample length

In this section, an estimation of the effect of sample length on the coupling loss is made. This is needed for the preparation of appropriate samples for AC loss measurements in the available set-up (section 4.4), which has a maximum sample length of only 15 cm.

In the coupling loss calculation in the beginning of this chapter, it was assumed that the coupling current extends to $x = \pm\infty$ along infinitely long bars. In cables of finite length, the coupling current forms a closed loop within the cable. The coupling current thus has to cross between the two cable halves. To calculate the coupling loss in such a sample, the 'cross-overs' of strands need to be considered. The simplest way to do this is to reduce each strand to a square wave that oscillates between the two cable halves. The y-coordinate of strand 1 is expressed as follows:

$$y_1(x) = \frac{W - w_s}{2} * \begin{cases} -1, & 0 < x < \frac{1}{2}\ell_t \\ 1, & \frac{1}{2}\ell_t < x < \ell_t \\ 0, & x = 0, \quad x = \frac{1}{2}\ell_t, \quad x = \ell_t \end{cases} \tag{4.34}$$

The square wave is periodic in the transposition length ℓ_t and has an amplitude of $W - w_s$, which equals the average displacement between the two cable halves. Each strand has the same geometry, but is shifted by ℓ_t/N in the x-direction compared to its neighbours. The square wave for each strand i is therefore:

$$y_i(x) = y_1\left(x - (i-1)\frac{\ell_t}{N}\right), \qquad i = 1, 2, \ldots, N \tag{4.35}$$

To calculate the coupling loss, the transverse currents between the strands need to be found. Let $J_i(x)$ be the current per unit length flowing from strand i to strand $i+1$. Considering a short loop of length Δx (see figure 4.4), Faraday's law results in the following equation:

$$(y_i - y_{i+1})\Delta x \frac{\mathrm{d}B_{ext}}{\mathrm{d}t} = \rho_a J_i(x + \Delta x) - \rho_a J_i(x) \tag{4.36}$$

where ρ_a is the unit-length resistance between adjacent strands.

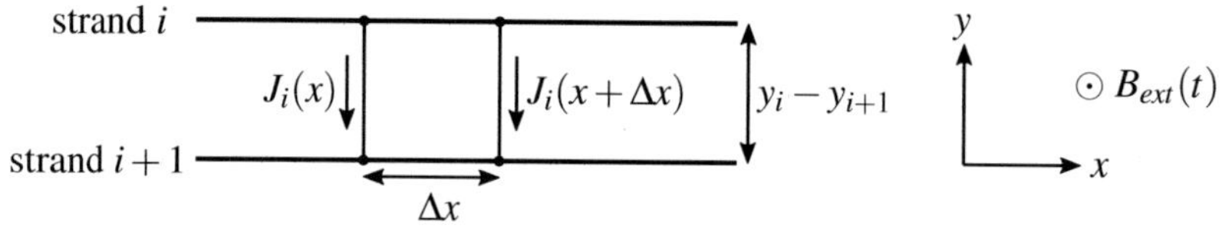

Figure 4.4: Induced transverse currents between neighbouring strands.

Dividing by Δx and taking the limit $\Delta x \to 0$:

$$(y_i - y_{i+1})\frac{\mathrm{d}B_{ext}}{\mathrm{d}t} = \rho_a \lim_{\Delta x \to 0} \frac{J_i(x+\Delta x) - J_i(x)}{\Delta x} = \rho_a \frac{\partial J_i}{\partial x} \tag{4.37}$$

The transverse current J_i can be found by integrating this equation with respect to x:

$$J_i = \frac{1}{\rho_a}\frac{\mathrm{d}B_{ext}}{\mathrm{d}t}\int (y_i(x) - y_{i+1}(x))\,\mathrm{d}x = \frac{1}{\rho_a}\frac{\mathrm{d}B_{ext}}{\mathrm{d}t}(Y_i(x) - Y_{i+1}(x) + a_i) \tag{4.38}$$

Here a_i is an integration constant and Y_i is the integral of y_i given by:

$$Y_1(x) = \frac{W - w_s}{2} * \begin{cases} -x, & 0 \le x \le \frac{1}{2}\ell_t \\ x - \ell_t, & \frac{1}{2}\ell_t \le x \le \ell_t \end{cases} \tag{4.39}$$

$$Y_i(x) = Y_1\left(x - (i-1)\frac{\ell_t}{N}\right) \tag{4.40}$$

The values of a_i can be found from the boundary conditions. Since no charge is accumulated in any strand, the integral of the transverse current flowing towards the strand J_{i-1} should equal that of the current away from the strand J_i. Therefore:

$$\int_{-\frac{1}{2}\ell}^{\frac{1}{2}\ell} (J_{i-1}(x) - J_i(x))\,\mathrm{d}x = 0 \tag{4.41}$$

where ℓ is the length of the cable. By filling in equation 4.38 into this boundary condition one finds:

$$\int_{-\frac{1}{2}\ell}^{\frac{1}{2}\ell} (Y_{i-1} - Y_i + a_{i-1} - Y_i + Y_{i+1} - a_i)\,\mathrm{d}x = 0 \tag{4.42}$$

$$\Rightarrow \quad a_i - a_{i-1} = \frac{1}{\ell}\int_{-\frac{1}{2}\ell}^{\frac{1}{2}\ell} (Y_{i-1} - 2Y_i + Y_{i+1})\,\mathrm{d}x \tag{4.43}$$

The first boundary condition gives $N-1$ independent equations, while there are N unknown constants. This second boundary condition is:

$$\sum_i J_i = 0 \tag{4.44}$$

This follows from the assumption that there are no rotating currents in the yz-plane. These currents can be induced only by the magnetic field in the x-direction, which is assumed to be zero. Inserting the current (equation 4.38) into the second boundary condition results in:

$$\sum_i a_i = 0 \tag{4.45}$$

The values of a_i can be found by solving a system of equations:

$$\begin{bmatrix} 1 & & & & -1 \\ -1 & 1 & & & \\ & \ddots & \ddots & & \\ & & -1 & 1 & \\ 1 & \dots & 1 & 1 & 1 \end{bmatrix} \begin{bmatrix} a_1 \\ a_2 \\ a_3 \\ \vdots \\ a_N \end{bmatrix} = \frac{1}{\ell}\int_{-\frac{1}{2}\ell}^{\frac{1}{2}\ell} \begin{bmatrix} Y_N - 2Y_1 + Y_2 \\ Y_1 - 2Y_2 + Y_3 \\ \vdots \\ Y_{N-2} - 2Y_{N-1} + Y_N \\ 0 \end{bmatrix} \mathrm{d}x \tag{4.46}$$

The last row of the matrix represents equation 4.45, while the rest follows from 4.43. The right hand side of the equation becomes zero for $\ell \to \infty$, leaving only the trivial solution $a_i = 0$. The right hand side becomes also zero if ℓ is an integer times the transposition length ℓ_t, because all functions Y_i are periodic in x. Therefore, cables with an integer number of transpositions have the same pattern of transverse currents as an infinitely long cable. For other values of ℓ the system can be solved using a linear algebra program such as Matlab.

Now the integrations constants a_i are known, the transverse currents can be computed from equation 4.38. The coupling loss power is found by integrating the unit length power $\rho_a J_i^2$ over the cable length and over all strands. The power per unit length is therefore:

$$P = \frac{1}{\ell} \sum_i \int_{-\frac{1}{2}\ell}^{\frac{1}{2}\ell} \rho_a J_i^2(x) \mathrm{d}x \tag{4.47}$$

The power as a function of sample length is shown in figure 4.5. The power is shown relative to that of an infinitely long cable. The curve slightly depends on the number of strands, but not on the other cable parameters. The sample length dependence of coupling loss for a round cable with a single layer of wires is shown as well. For such a cable an exact expression was derived by Ries and Takács [117]:

$$\frac{P}{P_\infty} = 1 - \left(\frac{\pi\ell}{\ell_t}\right)^{-2} \sin^2\left(\frac{\pi\ell}{\ell_t}\right) \tag{4.48}$$

As seen from the figure, the sample length dependence in a Roebel cable is similar to that of a single layer round cable, especially for higher numbers of strands.

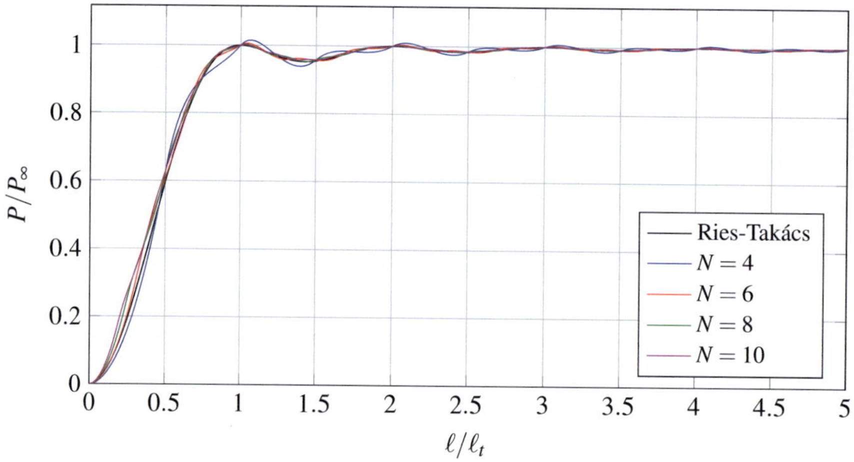

Figure 4.5: Coupling loss power per unit length as a function of cable length ℓ for different number of strands N. The result for a round cable derived by Ries and Takács [117] is shown for comparison.

4.3 Numerical calculation of AC loss with variable inter-strand resistance

In the previous section, the coupling losses were calculated assuming full penetration of the magnetic field. However, induced coupling currents interact with superconductor magnetization currents, partially shielding each other magnetically. For an accurate calculation of AC loss, it is thus beneficial to calculate both types of induced currents at the same time. In this section, a numerical method is proposed for computation of the combined hysteresis and coupling losses in a partially coupled Roebel cable.

A numerical method for calculating induced currents in a single superconducting strip was proposed by Brandt [118]. In this work, a superconducting tape is subdivided in elements with a non-linear resistivity. Using Maxwell's equations, a system of differential equation for the current in each element is derived. This system is numerically integrated to find the induced currents. The model described here follows a similar approach, but extends the geometry to two parallel stacks of superconducting strips. This geometry represents a Roebel cable in the limit of infinite transposition length. First, the cable is divided in a number of elements with uniform current. Then, an expression for the magnetic vector potential as a function of the current distribution is derived. The electric field is related to the vector potential by a time derivative. Together with an $E(J)$ relation this fully describes the current distribution in the cable. The resulting equation for the current distribution is numerically integrated to find the current as a function of time. Finally the dissipation is calculated from the product of current density and electric field.

4.3.1 Equation for the current distribution in a Roebel cable

The numerical approach is based on a cross-section of a Roebel cable extended to infinite length. This representation is two-dimensional, and therefore the altered current pattern at cross-overs is not taken into account. The cross-section is divided in N_{el} infinitely long sheets (figure 4.6). The elements are numbered $j = 1, 2, \ldots, N_{el}$ and each carry a uniform sheet current density J_j in the x-direction. Each element j is bordered on the left by $(y_{L,j}, z_j)$ and on the right by $(y_{R,j}, z_j)$. A time dependent external magnetic field $B_{ext}(t)$ is applied with components in both the y and the z-direction.

The magnetic vector potential at a distance r of an infinite wire carrying I is given by:

$$\mathbf{A} = -\frac{\mu_0 \mathbf{I}}{2\pi} \ln(r) \tag{4.49}$$

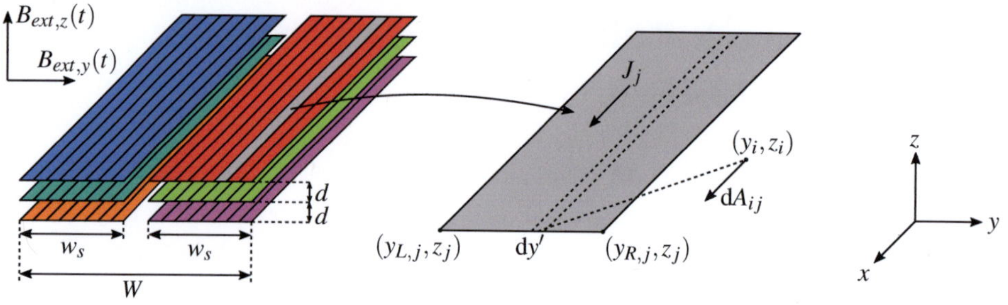

Figure 4.6: Representation of the Roebel cable for numerical calculations of AC loss. The cable is divided in infinitely long sheets carrying a sheet current density J_j in the x-direction. A time-dependent external field B_{ext} is applied with components in both the y and the z-direction.

The vector potential at (y_i, z_i) of a narrow sheet of width $\mathrm{d}y'$ carrying $J_j\mathrm{d}y'$ is therefore:

$$dA_{ij} = -\frac{\mu_0 J_j}{2\pi} \ln\left(\sqrt{(y_i - y')^2 + (z_i - z_j)^2}\right) \mathrm{d}y' \tag{4.50}$$

$$= -\frac{\mu_0 J_j}{4\pi} \ln\left((y_i - y')^2 + (z_i - z_j)^2\right) \mathrm{d}y' \tag{4.51}$$

The vector potential A_{ij} of element j is found by integration over the width of the element:

$$A_{ij} = -\frac{\mu_0 J_j}{4\pi} \int_{y_{L,j}}^{y_{R,j}} \ln\left((y_i - y')^2 + (z_i - z_j)^2\right) \mathrm{d}y' \tag{4.52}$$

To obtain the vector potential of the entire cable A_i at (y_i, z_i), the contributions of all elements are summed:

$$A_i = -\frac{\mu_0}{4\pi} \sum_j J_j \int_{y_{L,j}}^{y_{R,j}} \ln\left((y_i - y')^2 + (z_i - z_j)^2\right) \mathrm{d}y' \tag{4.53}$$

This can be expressed in the current J_j and a geometrical factor K_{ij}:

$$A_i = \frac{\mu_0}{2\pi} \sum_j K_{ij} J_j \tag{4.54}$$

where

$$K_{ij} = -\frac{1}{2} \int_{y_{L,j}}^{y_{R,j}} \ln\left((y_i - y')^2 + (z_i - z_j)^2\right) \mathrm{d}y' \tag{4.55}$$

$$= \begin{cases} (y_i - y_{R,j})\left(\ln|y_i - y_{R,j}| - 1\right) - (y_i - y_{L,j})\left(\ln|y_i - y_{L,j}| - 1\right) & z_i = z_j \\ \left[u\left(\frac{1}{2}\ln\left(u^2 + (z_i - z_j)^2\right) - 1\right) + (z_i - z_j)\tan^{-1}\left(\frac{u}{z_i - z_j}\right)\right]_{u = y_i - y_{L,j}}^{y_i - y_{R,j}} & z_i \neq z_j \end{cases}$$

The evaluation of the integral can be found in appendix B. Since K depends on the geometry only, it needs to be computed only once, and its evaluation is not time critical. The vector potential of the external magnetic field is given by:

$$A_{ext,x}(y,z) = -yB_{ext,z} + zB_{ext,y} \tag{4.56}$$

The total vector potential at (y_i, z_i) is thus given by:

$$A(y_i,z_i) = -y_iB_{ext,z} + z_iB_{ext,y} + \frac{\mu_0}{2\pi}\sum_j K_{ij}J_j \tag{4.57}$$

The magnetic vector potential and the electric field can be related using the following form of Faraday's law [119, p. 417]:

$$\mathbf{E} = -\nabla V - \frac{\partial \mathbf{A}}{\partial t} \tag{4.58}$$

Inserting the vector potentials and an external electric field in the x-direction described by $V = -xE_{ext,x}$ gives:

$$E_i(J) = -\frac{\mathrm{d}V}{\mathrm{d}x} - \frac{\partial}{\partial t}(A_i + A_{ext,x}(y_i,z_i)) \tag{4.59}$$

$$= E_{ext,x} - \frac{\mu_0}{2\pi}\sum_j K_{ij}\dot{J}_j + y_i\dot{B}_{ext,z} - z_i\dot{B}_{ext,y} \tag{4.60}$$

in which the dots denote the time derivative. This equation can be solved for $\dot{J}_i$ by inverting the matrix multiplication of K and J:

$$\sum_j K_{ij}\dot{J}_j = \frac{2\pi}{\mu_0}\left(E_{ext,x} - E_i(J) + y_i\dot{B}_{ext,z} - z_i\dot{B}_{ext,y}\right) \tag{4.61}$$

$$\Rightarrow \dot{J}_i = \frac{2\pi}{\mu_0}\sum_j K_{ij}^{-1}\left(E_{ext,x} - E_j(J) + y_j\dot{B}_{ext,z} - z_j\dot{B}_{ext,y}\right) \tag{4.62}$$

where K_{ij}^{-1} are the elements of the inverted matrix K^{-1}. Together with an $E(J)$ relation, e.g. the well-known power law, this differential equation fully describes the current distribution. The equation can be integrated using an explicit solver. In this case, Matlab's 'ode15s' function[1] was used. The right hand side of equation 4.62 is evaluated using matrix multiplications, which can be computed rapidly in Matlab.

[1] `https://mathworks.com/help/matlab/ref/ode15s.html`

4.3.2 Electric field as a function of the current distribution

In order to evaluate the right hand side of the current equation (equation 4.62), it is necessary to find the electric field as function of the current distribution. Here, a method to calculate the electric field considering the superconductor properties (J_c, n) and the effective resistance r_e is proposed.

The electric field in a superconductor is commonly described using a power law [120, p. 401]:

$$E(J) = E_c \left(\frac{J}{J_c(B_y, B_z)} \right)^n \tag{4.63}$$

In this equation, $E_c = 10^{-4}$ V/m is the critical electric field and n the non-linearity index usually in the range 20-50. The critical sheet current density J_c has units of A/m and may depend on both components of the local magnetic field B_y and B_z. Inserting the power law in equation 4.62 would represent the case of fully coupled strands, since there is no effect on the electric field for a screening current flowing in one strand and then flowing back in another.

Partial coupling is introduced by adding another term to the electric field proportional to the coupling current.

$$E(J) = E_c \left(\frac{J}{J_c(B_y, B_z)} \right)^n + E_{coupling} \tag{4.64}$$

To find the coupling current, it is assumed that the total current in each strand is a superposition of superconductor magnetization current and coupling current, and that the transport current is zero. Since the magnetization current returns within the same strand, its integral over the strand width is zero. Therefore, the coupling current in a strand can be found by summing the current carried by each element belonging to that strand. Separating the contributions to the current in this way has a number of drawbacks which will be discussed in section 4.3.4.

As shown in section 4.2.1, the coupling current experiences an effective resistance per unit length of:

$$r_e = \frac{2\rho_a N}{\ell_t^2} \tag{4.65}$$

Since each cable half contains $N/2$ strands, the electric field contribution related to coupling currents is:

$$E_{coupling} = \frac{N}{2} r_e I_{coupling} \tag{4.66}$$

where $I_{coupling}$ is the coupling current in one strand. In this way, the total coupling current experiences an effective resistance of r_e, because it is carried by $N/2$ parallel strands.

In coated conductors, the critical current density depends strongly on the magnetic field. Implementing this field dependency in the calculation can improve the accuracy. The y and z components of the magnetic field can be found by applying $\mathbf{B} = \nabla \times \mathbf{A}$ to equation 4.57:

$$B_y(y_i, z_i) = B_{ext,y} + \frac{\mu_0}{2\pi} \sum_j J_j \frac{\partial K_{ij}}{\partial z_i} \tag{4.67}$$

$$B_z(y_i, z_i) = B_{ext,z} - \frac{\mu_0}{2\pi} \sum_j J_j \frac{\partial K_{ij}}{\partial y_i} \tag{4.68}$$

The derivatives of K_{ij} can be found in appendix B. These formulas are used to calculate the magnetic field at each time step and adjust the critical current accordingly. This is done by interpolation in measured values of the critical current in magnetic field, which will be discussed in section 4.4.

4.3.3 Calculation of AC loss and hysteresis loops

The unit-width current and electric field in each element have been calculated using equations 4.62 and 4.64. From these values, the dissipated power in each element can be calculated. By integrating the power over an entire cycle, the AC loss is computed.

The dissipated power per unit width in element j equals the product of current per unit width J_j and the electric field E_j. The dissipation in the entire cable equals the sum of dissipation in all elements:

$$P = \sum_j J_j E_j w_j \tag{4.69}$$

To find the loss per cycle Q, the power P is integrated over a time $1/f$. Transient effects can occur at the beginning of the simulation. To avoid influence on the calculated loss, the current distribution is calculated over four entire periods, while Q is taken from the last period only.

$$Q = \int_{3/f}^{4/f} P(t)\mathrm{d}t \tag{4.70}$$

Alternatively, the loss per cycle can be computed from the magnetic hysteresis loop. Assuming a zero transport current, the magnetic moment per unit length is:

$$\mathbf{m} = \sum_j J_j w_j \left(z_j \hat{\mathbf{y}} - y_j \hat{\mathbf{z}}\right) \tag{4.71}$$

where $\hat{\mathbf{y}}$ and $\hat{\mathbf{z}}$ are unit vectors in the y- and z-directions. The work done on the sample during a small change in magnetic field is $-\mathbf{m}\cdot\mathbf{dB_{ext}}$. To find Q the work is integrated over an entire cycle:

$$Q = -\oint \mathbf{m}\cdot\mathbf{dB_{ext}} = -\oint m_y \mathrm{d}B_{ext,y} - \oint m_z \mathrm{d}B_{ext,z} \tag{4.72}$$

This integral equals the area enclosed by the hysteresis loop $m(B_{ext})$.

Several example calculations of AC loss in a sinusoidal field using the cable parameters in table 4.1 are made. These cable parameters are typical of the samples investigated in chapter 3, where the inter-strand resistance was studied. An inter-strand resistance (ISR) of $\rho_a = 0.63\ \mu\Omega\text{m}$ is used, which is between the values for soldered and pressed cables. The critical current in these examples is assumed to be independent of magnetic field. The external field points in the z-direction, perpendicular the the wide surface of the strands:

$$B_{ext,y} = 0 \tag{4.73}$$

$$B_{ext,z} = B_0 \sin(2\pi f t) \tag{4.74}$$

Table 4.1: Cable parameters used for example calculations.

	symbol	value	unit
Cable width	W	12.0	mm
Strand width	w_s, w_c	5.5	mm
Strand thickness	d	0.10	mm
Transposition length	ℓ_t	126	mm
No. of strands	N	10	-
Critical current density	J_c	25	A/mm
n-value	n	30	-
Adjacent ISR	ρ_a	0.63	μΩm

Figure 4.7 shows the numerically calculated AC loss of a single strand with different n-values. The AC loss was calculated using equation 4.70, which was derived in the previous section. For comparison, the AC loss of an ideally superconducting thin strip with an n-value approaching infinity is shown as well. For such strip, there is an exact expression which was derived by Halse [121] and later by Brandt and Indenbom [122]:

$$Q = \frac{2\mu_0 J_c^2 w^2}{\pi}\left\{\ln\left(\cosh\left(\frac{\pi H_0}{J_c}\right)\right) - \frac{\pi H_0}{2J_c}\tanh\left(\frac{\pi H_0}{J_c}\right)\right\} \tag{4.75}$$

In this equation $H_0 = B_0/\mu_0$ is the peak external field during the cycle. With increasing n-value, the numerically computed AC loss approaches that of the ideally superconducting strip.

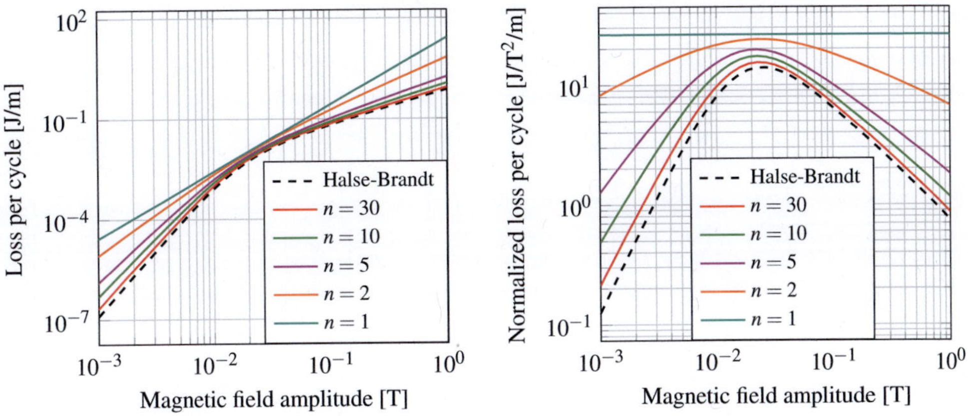

Figure 4.7: Calculated AC loss in a single tape of width $w = 5.5$ mm and with a critical current density $J_c = 25$ A/mm. The AC loss was calculated for different n-values using the proposed numerical method. The AC loss of an ideally superconducting strip calculated using the Halse-Brandt formula (equation 4.75) is shown for comparison. In the right figure the loss is normalized to B_0^2 to make the difference between results more visible.

Figure 4.8 shows hysteresis loops (equation 4.71) of the 10-strand cable for different field amplitudes and frequencies. The typical diamagnetic behaviour of superconductors can be observed. The magnetization saturates with increasing field amplitude when induced currents extend the whole cable, and no additional current can flow in the superconducting state. With increasing frequency, the area of the loop increases. This can be ascribed to coupling loss, although the hysteresis loss of a power-law superconductor is slightly frequency dependent itself [123].

In figure 4.9, the numerically computed loss per cycle is compared to the coupling loss computed using the approximation of two infinitely long bars (section 4.2, equation 4.17). The numeric calculation includes hysteresis loss in the superconducting layer itself, while the coupling loss formula calculates coupling loss only. The screening currents induced in the superconducting layer have infinite time constant, and therefore the hysteresis loss is practically frequency independent. For this reason, the hysteresis losses dominate at low frequencies, where coupling losses are small. The coupling losses make up the largest share of the total loss at frequencies close to $2\pi\tau$ and at low field amplitudes. At high field amplitudes, hysteresis loss dominates because it increases more strongly with field amplitude than coupling loss.

Interestingly, the coupling loss formula sometimes exceeds the numerically computed total loss. This is possible because screening currents expel part of the magnetic flux from the cable, decreasing the driving force of coupling currents. Such screening currents are taken into account in the numeric calculation, but not in the derivation of the coupling loss formula. Although this effect could lead to lower loss in a single cable, it is unlikely to do so in a coil, because the expelled flux would increase loss in nearby windings. Also, the coupling loss formula does not take into account an increase in effective resistance that occurs when the induced current exceeds the critical current. This can occur at very high field amplitudes.

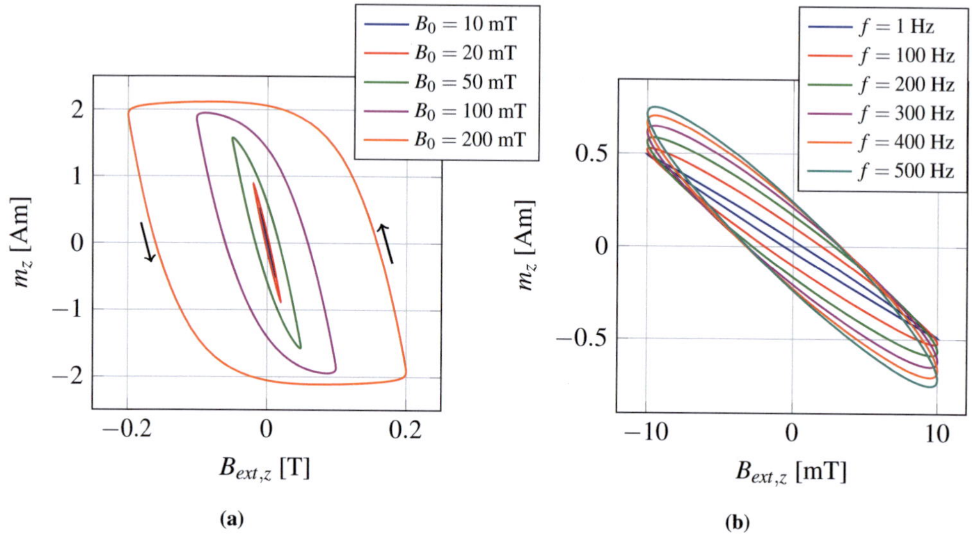

Figure 4.8: Hysteresis loops of a Roebel cable with (a) different field amplitudes ($f = 1$ Hz) and (b) different frequencies ($B_0 = 10$ mT). $B_{ext,z}$ is the external magnetic field in the z-direction, which changes sinusoidally in time. m_z is the magnetic moment per unit length of the cable in the z-direction, and was calculated using equation 4.71. The cable parameters used for these calculations are listed in table 4.1.

4.3.4 Limitations of the calculation method

Using the proposed numerical method, the AC losses of a Roebel cable with variable inter-strand resistance can be computed quickly. Numerical integration of equation 4.62 can be implemented in a program such as Matlab, and the calculation is fast. However, the model is based on a cross-section of the cable extended to infinite length, and does not perfectly represent the three-dimensional structure of a Roebel cable. This leads to a number of limitations:

Since only longitudinal currents along the cable are calculated, the precise distribution of transverse current between strands is not known. For this reason, the coupling current is separated from superconducting magnetization assuming a uniform distribution of coupling currents in each strand (see section 4.3.2). In reality, the distribution of coupling currents is not necessarily uniform. Currents between strands are hindered by the highly resistive substrate and will flow mostly in the copper coated edges. Such effects are not taken into account in the calculation.

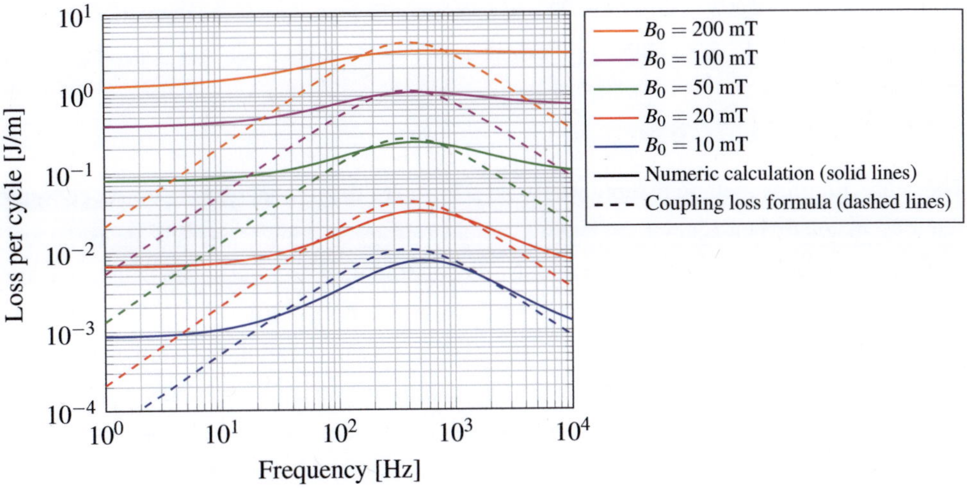

Figure 4.9: Comparison of the numeric calculation of AC loss and the coupling loss formula for a Roebel cable. The numeric calculation includes hysteresis loss in the superconducting layer itself, while the coupling loss formula calculates coupling loss only. The used cable parameters are listed in table 4.1.

Secondly, the method does not consider the transposition of strands. The influence of transverse currents at cross-overs on AC loss is therefore not computed. Arguably, this effect is negligible because the transposition length of the cable is much longer than the cable width, so that transverse currents make only a very small contribution to the AC loss [124].

Since each strand is modelled as an infinitely thin superconducting layer, currents and losses in the metal layers are not calculated.

Finally, the loss is calculated on a single cable in free space, ignoring interaction with nearby windings. This is not fully representative for the potential applications of Roebel cables, which are all in the form of coils.

All limitations could be overcome using a fully three-dimensional model of the cable or device, at the cost of much higher complexity and longer computation times. Such a three-dimensional was developed by Zermeño et al. [125], though it does not consider partial coupling of strands.

4.4 AC loss characterization using the calibration-free method

To validate the calculation method for coupling losses, the AC loss was measured in several cables with different levels of inter-strand resistance. The calibration-free method was used to measure the AC magnetization loss. The measurements were done over the widest possible range of field frequencies, to obtain as much as possible data about frequency dependent effects.

4.4.1 Calibration-free method

The calibration-free method determines AC loss by measuring the increase in power consumption of a magnet when a superconducting sample is inserted. Unlike AC loss measurements using pickup coils or calorimetric methods, no calibration is required. The calibration-free method was first developed at the Slovak Academy of Sciences [126].

The power dissipation in a magnet is given by the product of in-phase components of the applied current and the voltage over the magnet. In principle, AC loss of any sample could be determined by comparing the dissipation with and without inserted sample. A difficulty is that the dissipation of the superconducting sample is often small compared to that of the magnet. The measurement signal is thus likely to be overwhelmed by resistive and inductive voltages of the magnet itself. To increase the sensitivity, an arrangement of coils as shown in figure 4.10 is used. The system consists of two identical racetrack coils wound from a braided copper wire with insulated filaments. One of the filaments is separated from the rest at the coil ends. This filament is used as a measurement coil, while all other filaments are used to transport the current. The measurement coil carries no current and therefore has no resistive voltage, but it retains the in-phase component due to dissipation in the sample. To compensate the out-of-phase (inductive) voltage, the two identical racetrack coils are connected in series, with the measurement wires connected in anti-series. The only remaining voltage on the measurement wire is due to magnetization of the sample.

In practice, the coils are not perfectly identical, and some weak signal remains even when no sample is inserted. This signal is compensated using two correction coils. The first correction coil is placed inside racetrack coil 1 and connected to a variable resistor. The correction coil acts as a variable load on racetrack coil 1, and is used to match the in-phase voltages over both racetrack coils. The out-of-phase voltage mismatch is corrected using the second correction coil, which is connected series with the measurement coils. This correction coil is inductively coupled to a racetrack coil by placing it close to the LN_2 container. By changing the orientation or distance to the racetrack coil, the level of coupling can be adjusted in such a way that the induced voltage in the correction coil cancels the out-of-phase voltage related to coil imperfections.

To find the dissipation in the sample, the component of the measurement signal with the same frequency and phase as the applied current must be extracted. This is usually done using a lock-in amplifier, which can extract the desired signal even with low signal-to-noise ratios. The measurements described in this work were done not using a separate lock-in amplifier, but using

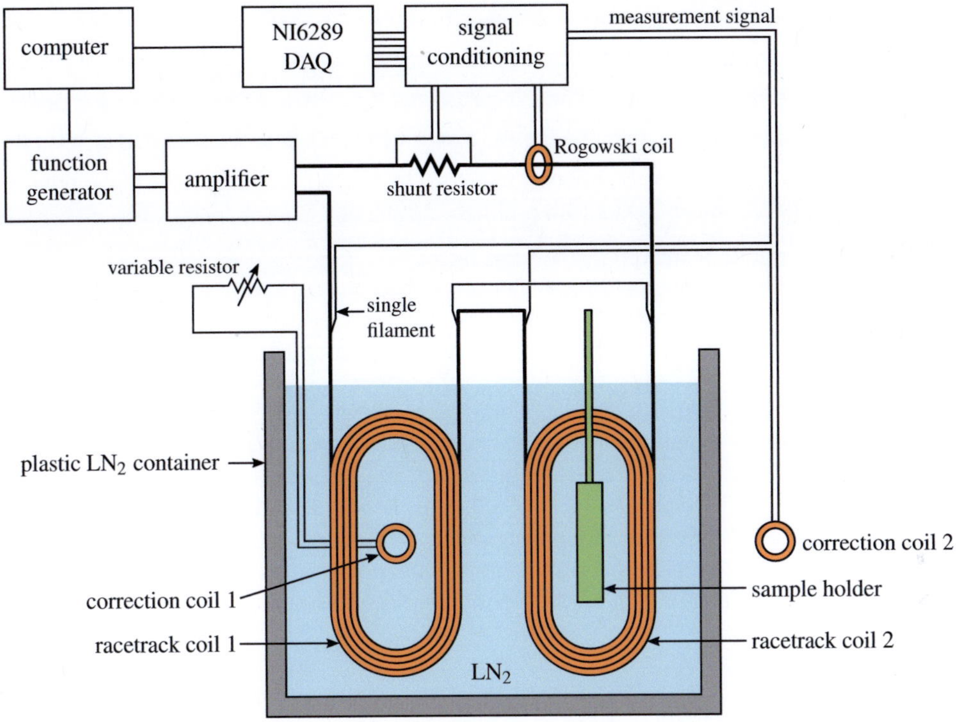

Figure 4.10: Scheme of the calibration-free AC loss measurement set-up

a general purpose data acquisition (DAQ) device and a virtual lock-in amplifier in LabVIEW. The reference signal for the lock-in amplifier is supplied by a Rogowski coil inductively coupled to the racetrack coils. This provides a reference signal shifted by 90° compared to the applied current. The amplitude of the reference signal is proportional to the frequency of the applied current, and can be insufficient at low frequencies. To enable measurements at the lowest amplitudes and frequencies, a shunt resistor is used, which provides a reference signal in phase with the applied current.

The NI6289 data acquisition device used for the measurement is a multiplexing device. The different channels are alternately scanned by a single analog-to-digital converter. This leads to a small time shift between reference and measurement signals. In this specific case, the sampling rate was 480 kHz and the reference and measurement signals were two channels apart, leading to a time shift of 4.17 μs. This time shift is insignificant at low frequencies, but causes a spurious phase shift of 3.07° at the highest measured frequency of 2048 Hz.

Figure 4.11 shows the effect such a phase shift can have on the hysteresis loop at a low amplitude of 5 mT and a high frequency of 2048 Hz. In this case, the magnetic moment has a large component in-phase and a small component out-of-phase with the magnetic field. The phase shift caused by time difference between measurement signals distorts the hysteresis loop. Part of the out-of-phase component is now in-phase and vice versa. If the in-phase component is large, this can have a strong effect on the AC loss which is equal to the enclosed area. In the case of figure 4.11, the AC loss is spuriously increased by a factor 2.36. Since the measurements signal in the calibration-free measurement is proportional to the magnetic moment, the measured AC loss will be wrongly increased by the same factor.

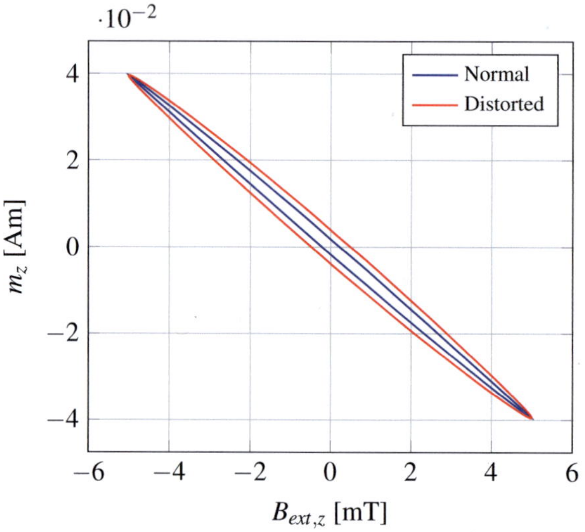

Figure 4.11: Calculated hysteresis loop with and without distortion by multiplexer lag at a frequency of 2048 Hz and a magnetic field amplitude of 5 mT. The area enclosed by this particular distorted hysteresis loop is increased by a factor 2.36.

All measurements described in this chapter were done with the multiplexing device. The error and its correction will be discussed further in section 4.4.3. For future measurements, it is recommended to use a data acquisition device with simultaneous sampling[2]. Using such a device there is no phase shift that can cause errors. At least, the phase shift should be corrected in LabVIEW.

4.4.2 Sample preparation and DC characterization

The Roebel cables made at ITEP usually are 12 mm wide and have a transposition length of 126, 226 or 300 mm. Since the racetrack coils can accommodate a 150 mm long sample, only a single transposition or less could be measured. For representative measurements of

[2] For example National Instruments PCI-6143

coupling loss it is recommended that the sample is at least several transposition lengths long (see section 4.2.2). Therefore, “miniature” Roebel cables with very short transposition lengths were prepared. These samples have a width of 4 mm and and a transposition length of 50 mm or 30 mm, so that three or five full transpositions can be measured in the existing set-up.

All samples were prepared using SCS4050-AP coated conductor from SuperPower. The conductor has a 50 μm thick Hastelloy C-276 substrate and is coated with 20 μm of copper on both sides. Roebel strands of 20 cm length were prepared from this wire by laser cutting. Unlike the punching tools with fixed geometry, laser cutting can be used to prepare short lengths of Roebel cable with any combination of cable parameters.

Four cables with six strands each were prepared. The Roebel cable parameters are listed in table 4.2. Cables 1 and 3 have a transposition length of 50 mm, while cables 2 and 4 have 30 mm. The strands of cables 1 and 2 were interconnected with In52Sn48 solder to attain a low inter-strand resistance. The procedure for preparing a soldered cable is the same as used for previous inter-strand resistance measurements and is described in section 3.5.2. Cables 3 and 4 were compressed in a purpose-made sample holder using plastic bolts (figure 4.12) to reach an intermediate inter-strand resistance. The precise pressure applied is unknown, but this value is not a precise predictor for the inter-strand resistance due to uncontrollable contact areas between strands (see chapter 3). The entire sample holder is made of non-magnetic and non-conducting materials, so that it does not influence the AC loss measurement. After the AC loss measurements, the strands of cables 3 and 4 were insulated and AC loss was measured one more time. Insulation was applied by disassembling the cable, adding polyimide tape to both sides of each strand, and then reassembling the cable. The insulation increased the strand thickness to 0.22 mm.

Table 4.2: Properties of the cable samples.

		Cable 1	Cable 2	Cable 3		Cable 4		unit
Type		soldered	soldered	pressed	insulated	pressed	insulated	
Cable width	W	4.0	4.0	4.0	4.0	4.0	4.0	mm
Strand width	w_s, w_c	1.9	1.9	1.9	1.9	1.9	1.9	mm
Strand thickness	d	0.10	0.10	0.10	0.22	0.10	0.22	mm
Transposition length	ℓ_t	50	30	50	50	30	30	mm
Sample length	ℓ	150	150	150	150	150	150	mm
No. of strands	N	6	6	6	6	6	6	-
Strand critical current	I_c	50.2	49.3	50.8	50.8	50.3	50.3	A
n-value	n	23.7	22.9	20.4	20.4	25.0	25.0	
Adjacent ISR	ρ_a	0.265	0.295	2.47	∞	2.39	∞	μΩm
Crossing ISR	ρ_c	1.07	1.21	$\gg \rho_a$	∞	$\gg \rho_a$	∞	μΩm

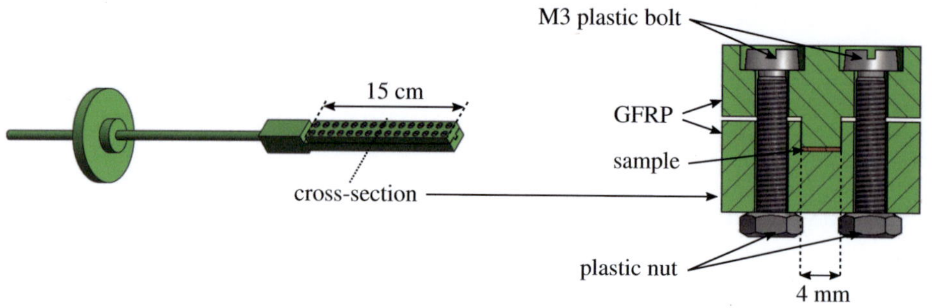

Figure 4.12: Sample holder for AC loss measurements.

Besides cable geometry, critical current and inter-strand resistance values are needed as input for the AC loss calculations, to which the measurements will be compared. The critical currents and *n*-values of all separate strands were measured in a liquid nitrogen bath before cable assembly. The average values for each cable are shown in table 4.2. For increased accuracy of the calculation it is beneficial to take into account that the critical current depends on the magnetic field magnitude and angle. For that purpose, the angle and field dependent critical current of a coated conductor sample was measured (figure 4.13). At low external fields, the self-field dominates and the measured critical currents are not representative of the local $J_c(B)$ values. The local $J_c(B)$ without self-field influence was extracted from the data using software developed by Zermeño et al. [125].

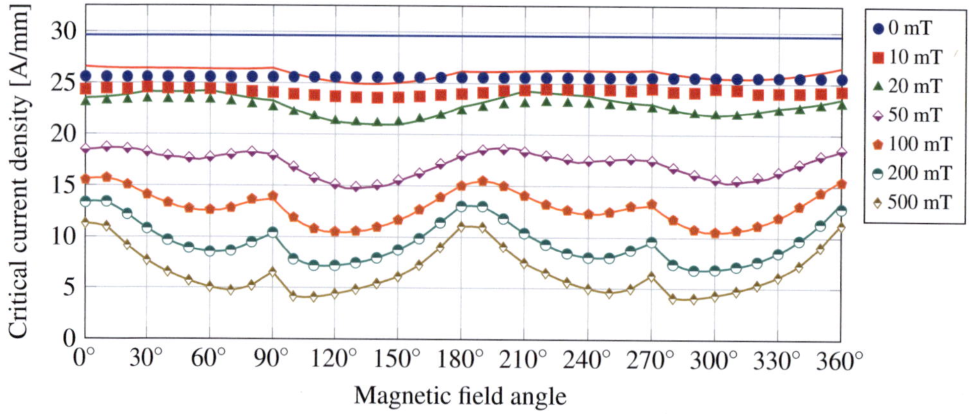

Figure 4.13: Critical current ($T = 77$ K) per unit width as a function of external magnetic field magnitude and angle measured on a sample of SuperPower SCS4050-AP conductor. An angle of 0° corresponds to a field perpendicular to the wide face. The markers show the critical current that was measured directly on the tape. The lines show the local critical current extracted from the measurement data.

After mounting the cable in the sample holder, the inter-strand resistance values ρ_a and ρ_c were determined using the method described in section 3.5. In short, this is done by applying a current between opposing strands, measuring the voltage profile, and resolving ρ_a and ρ_c by curve fitting. The measurement was repeated through three thermal cycles for each cable. The results are shown in figure 4.14. The ρ_a and ρ_c values of the soldered cables are relatively stable with thermal cycling. The ρ_a values of the pressed cables are unfortunately not as stable with thermal cycling as the soldered cables. This is likely due to a change in transverse stress resulting from differential thermal expansion. To keep further change of ρ_a values to a minimum, these samples were stored at 77 K between AC loss measurements.

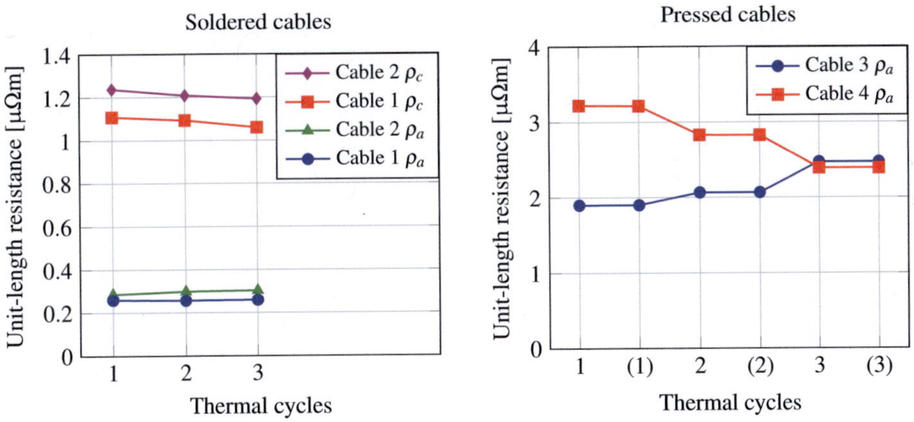

Figure 4.14: Inter-strand resistance measurements ($T = 77$ K) in the soldered and pressed cable samples with thermal cycling. The inter-strand resistance of the pressed cables was measured twice after each thermal cycle.

4.4.3 Results

Magnetization AC loss was measured in a sinusoidal transverse field perpendicular to the wide face of the cable. The measurements were done with field frequencies ranging from 1 to 2048 Hz, doubling with each step. The maximum field amplitude is 70 mT for frequencies up to 64 Hz, but decreases for higher frequencies because of high inductive voltages. When both field amplitude and frequency are low, the measurement signal is very weak and measurements become unreliable. The lowest measurable field amplitude therefore increases with decreasing frequencies.

The measurements and calculations for all samples are shown in figures 4.15 to 4.17, ordered to increasing inter-strand resistance (see table 4.2). The solid lines are calculated numerically (equation 4.70) with the corresponding ρ_a value. The dashed lines equal the sum of the numerical AC loss of an insulated cable and coupling loss following equation 4.17. The dashed lines thus do not take into account inter-action between superconductor magnetization and coupling currents. For all calculations, the last measured inter-strand resistance value ρ_a was used (see figure 4.14). The open markers represent the directly measured AC loss. The closed markers, which largely overlap with the open markers, were corrected for the error due to multiplexed sampling (see section 4.4.1). This was done by dividing the measured loss by the calculated increase in area of the hysteresis loop. This correction factor depends on field frequency, amplitude and loss level, and was computed for each measurement point separately.

Figure 4.15 shows the AC loss of the soldered cables. The AC loss matches well with the numerical calculation at amplitudes of 5 mT and higher. At low amplitudes and high frequency, the calculation somewhat overestimates the loss. The pressed cables (figure 4.16) have a higher inter-strand resistance and therefore a lower time constant. This shifts the peak in coupling losses to higher frequencies, which can be seen in both the measurement and the calculation. The calculation is again the most accurate for field amplitudes above 5 mT. In all four cables, the dashed lines exceed the measured loss, except for the lowest amplitudes and frequencies. This supports the use of equation 4.17 as an upper limit for the coupling loss.

The AC loss of the cables with insulated strands is shown in figure 4.17. At low amplitudes and high frequencies, the measured AC loss is much higher than the calculation. However, the difference is resolved when the measurements are corrected for the error resulting from multiplexing. The correction factor is larger than in the coupled cables, because the AC losses are smaller compared to spurious contribution from the in-phase component of the magnetic moment. This underlines the need to use simultaneous sampling of reference and measurement signal. The numerical calculation predicts that, at low amplitudes, the AC loss decreases slightly with frequency. This is because, at high frequencies, the external field penetrates less far into the conductor leading to lower AC loss [123, 127]. However, the spread between measurement points is too large to confirm this effect in the Roebel cables.

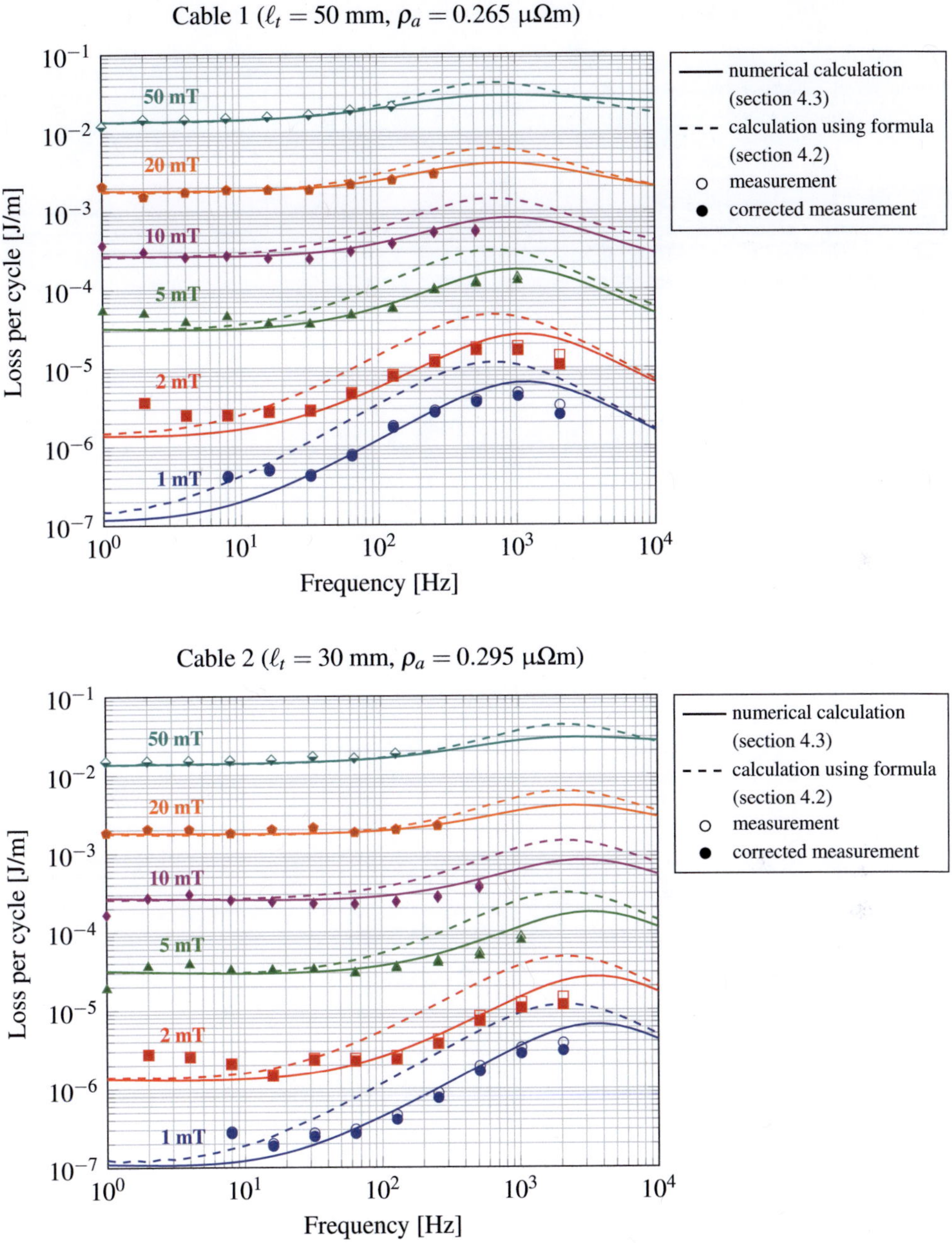

Figure 4.15: AC loss per cycle per unit length in soldered cables 1 and 2. Solid line: numerical calculation. Dashed line: sum of the numerical calculation of an insulated cable and coupling loss following equation 4.17. Open markers: AC loss measurement. Closed markers: AC loss measurement corrected for multiplexer lag. The last measured ρ_a value, which was used for both calculations methods, is shown above each diagram.

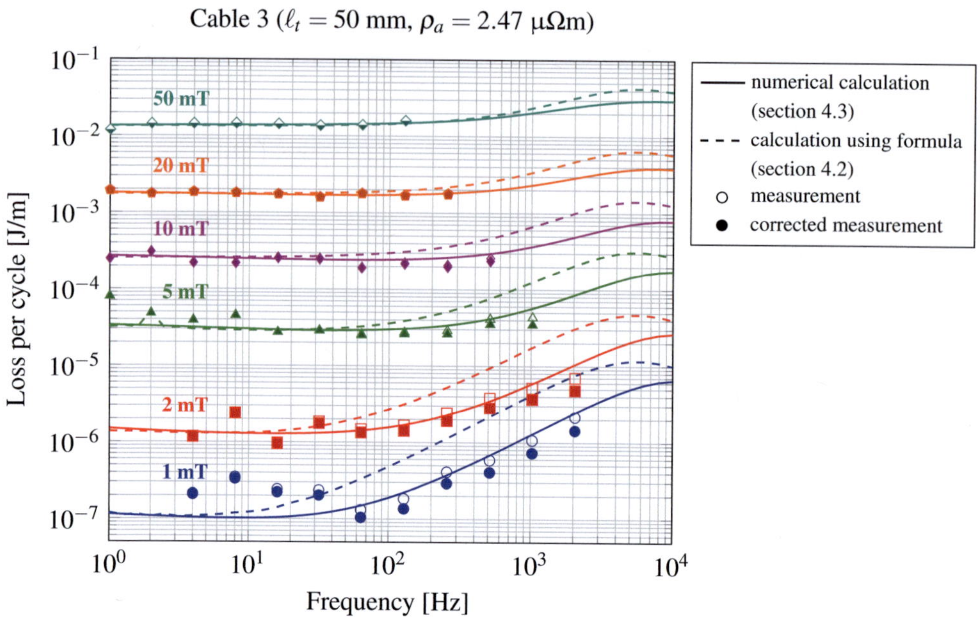

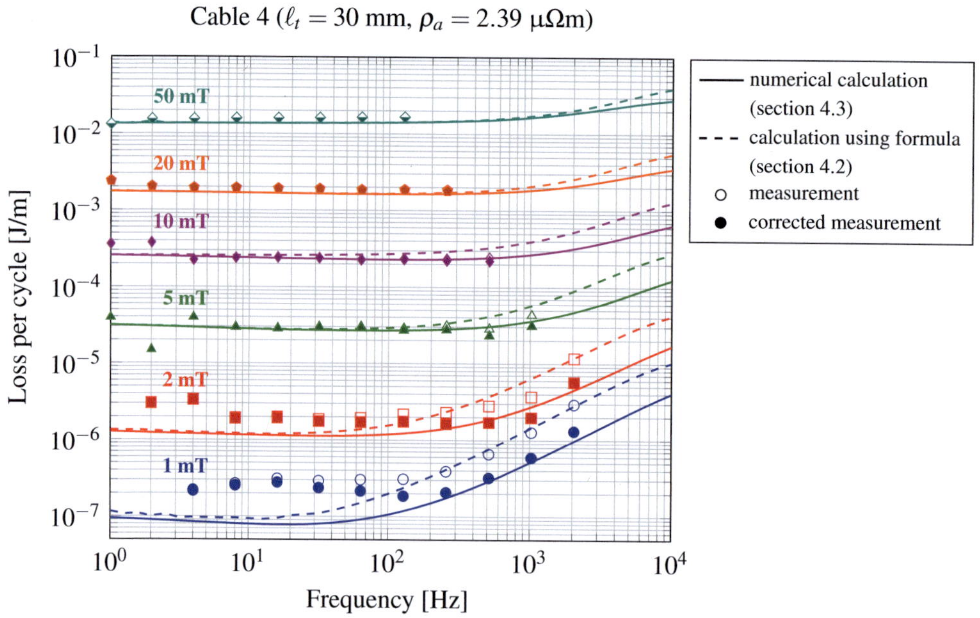

Figure 4.16: AC loss per cycle per unit length in pressed cables 3 and 4. Solid line: numerical calculation. Dashed line: sum of the numerical calculation of an insulated cable and coupling loss following equation 4.17. Open markers: AC loss measurement. Closed markers: AC loss measurement corrected for multiplexer lag. The last measured ρ_a value, which was used for both calculations methods, is shown above each diagram.

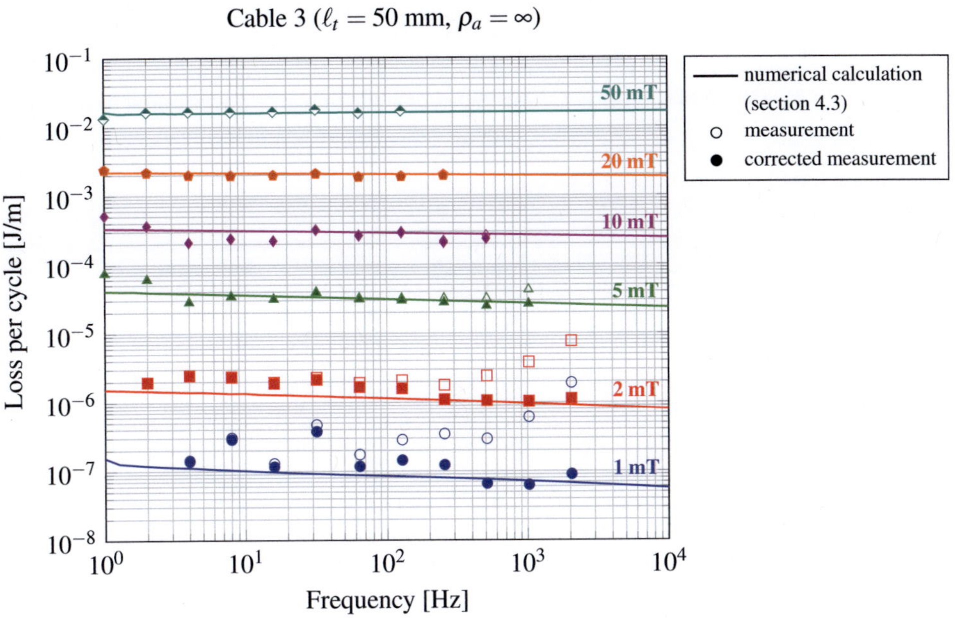

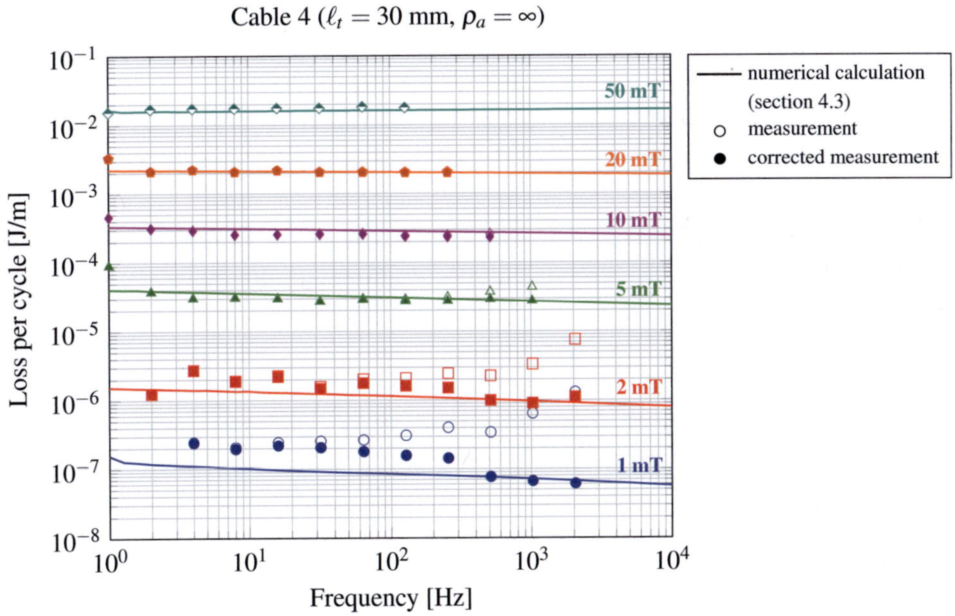

Figure 4.17: AC loss per cycle per unit length in cables 3 and 4 after insulating the strands. Solid line: numerical calculation. Open markers: AC loss measurement. Closed markers: AC loss measurement corrected for multiplexer lag.

4.5 Conclusion and recommendations

The AC loss of Roebel cables with variable inter-strand resistance in a sinusoidal field perpendicular to the wide face was investigated. Using an approximation of the cable with two parallel bars, the coupling loss can be estimated by:

$$Q = \frac{(W - w_s)^2 B_0^2 \pi^2 f}{r_e (1 + (2\pi f \tau)^2)} \tag{4.76}$$

where r_e is the effective resistance per unit length and τ is a time constant. Based on this equation, the coupling loss per cycle is expected to be proportional to the frequency if $f \ll (2\pi\tau)^{-1}$ and inversely proportional if $f \gg (2\pi\tau)^{-1}$. The time constant τ is inversely proportional to the effective resistance r_e.

A numerical calculation method of AC loss based on a cross-section was also proposed. This method calculates both superconductor magnetization and coupling currents, and takes into account interaction between both effects. The numerical calculation predicts a lower level of coupling loss than the parallel bar approximation, because superconductor magnetization partially expels the flux that induces coupling currents.

The magnetization AC loss in several Roebel cables was measured over a wide range of frequencies (1-2048 Hz) using the calibration-free method at 77 K. Roebel cables with different levels of inter-strand resistance were prepared by soldering the strands together, applying pressure to the cable or insulating the strands. All cables had a short transposition length of 50 mm or 30 mm, so that at least three full transpositions could be measured. The measured AC loss matches the numerically computed loss well for amplitudes of 5 mT and higher. Except for the lowest frequencies and amplitudes, the coupling loss equation based on parallel bars exceeds the measured coupling loss.

A general purpose DAQ device with a virtual lock-in amplifier in LabVIEW can be an alternative to a separate lock-in amplifier. However, it is recommended to use a DAQ with simultaneous sampling of the reference and the measurement signal. The use of a multiplexed DAQ leads to a spurious increase in measured AC loss at high frequencies.

Above findings can help to estimate coupling loss induced by perpendicular fields in applications with *RE*BCO Roebel cables, and to select cable parameters that lead to acceptable loss levels. It is difficult to recommend a specific inter-strand resistance value in general, because the level of AC loss depends not only on the cable properties but also on the operating conditions such as the current, ramp rates, magnetic field magnitude and direction, temperature etc. Further, the acceptable heat load from AC loss may differ. A more precise analysis of AC loss can be done together with development of the device, as more of these conditions are known.

Chapter 5

Current redistribution and effect on stability

5.1 Introduction

The temperature in a superconducting wire can locally increase for a number of reasons including eddy currents, wire movements, local defects, poorly soldered joints and heat leaks. A local increase in temperature can lead to the formation of a normal zone, where the electric field is non-zero. The temperature in this normal zone can start to rise very rapidly making a return to the superconducting state impossible unless the current is stopped. Such a thermal runaway, called a quench, is undesirable because it makes the superconducting device temporarily unusable and can lead to overheating damage.

Large superconducting devices are usually equipped with quench protection systems which stop the current once a normal zone is detected. Additionally, the quench can be prevented from starting by improving the stability of the superconducting wire. The resistivity of superconductors in the normal state is rather high. For this reason, most practical superconductors are stabilized with a highly conductive metal such as copper, silver or aluminium. The stabilization provides an alternative path for the current if a normal zone develops, reducing Joule heating. The stabilization also helps to remove heat from the normal zone by thermal conduction. Stabilized wires have a higher chance of recovery from local disturbances, and, in case of a quench, the temperature rises more slowly.

In some cases, the stability can be improved further by providing alternative paths for the current outside of the wire. A notable development have been no-insulation HTS pancake coils, which are wound without turn-to-turn insulation [11, 128]. The current can bypass a normal zone through the contact between windings, reducing further Joule heating. Such coils have been shown to be more stable than insulated coils [129–131]. In LTS Rutherford cables, the parallel strands provide alternative current paths to each other, improving stability if the inter-strand resistance is sufficiently low [50, 132]. A similar effect in *RE*BCO Roebel cables was speculated about by Schuller et al. [112], but was not further investigated. J. van Nugteren predicted an increased stability of Roebel cables due to current redistribution using a numerical model [133, p. 185]. Until now, however, no experimental validation of such improved stability exists.

In this chapter, the effect of current redistribution in Roebel cables on stability is investigated. As a measure of stability, the minimum quench energy (MQE) is used. The MQE is defined as the minimum amount of energy deposited in one of the strands that causes a quench of the entire cable. The main questions are how the MQE depends on the level of coupling between the strands, and in which way stability can be optimized. In section 5.2, the classical theory of minimum propagating zones is discussed. It can be used to predict a MQE value for single superconducting wires with a constant current distribution. In section 5.3, a set of equations for the current and temperature distribution in a superconducting cable is derived. The equations are numerically solved using a finite-difference approximation, while taking into account the temperature-dependent properties. In this way, the response of the cable to an external heat pulse can be simulated. In order to understand the effect of inter-strand resistance on the MQE, a large number of MQE calculations were done with different parameters. These are discussed in section 5.5. Section 5.6 describes MQE measurements on a Roebel cable which were done in quasi-adiabatic conditions. In this experiment, a quench heater is placed on one of the strands. Heat pulses of increasing magnitude are applied, until the cable quenches. The experimental data is compared to the calculations.

5.2 Quench initiation

5.2.1 Transition to the resistive state

For calculations of quench dynamics, it is necessary to have an accurate description of the superconducting wire when it carries a current larger than the critical current I_c. The simplest model is the current sharing model, which assumes that all current exceeding the critical current is carried by the metallic part of the wire. This leads to an electric field of

$$E = \begin{cases} 0 & I \leq I_c \\ \frac{\rho_e}{A}(I - I_c) & I \geq I_c \end{cases} \tag{5.1}$$

In this equation, ρ_e is the effective resistivity with contributions of all metallic layers, and A is the cross-sectional area of the conductor. The electric field as function of current is shown in figure 5.1. Since the current in the superconducting part of the conductor never exceeds the critical current, there is an abrupt change to resistive behaviour at $I = I_c$. In practice, the transition is not that abrupt, and the electric field near the critical current is better described by an empirical power law [120, p. 401]:

$$E = E_c \left(\frac{I}{I_c} \right)^n \tag{5.2}$$

where $E_c = 10^{-4}$ V/m is the critical electric field and n is a non-linearity index. The power law is however not suitable for currents much larger than I_c as it does not consider the resistive behaviour of the metallic layer.

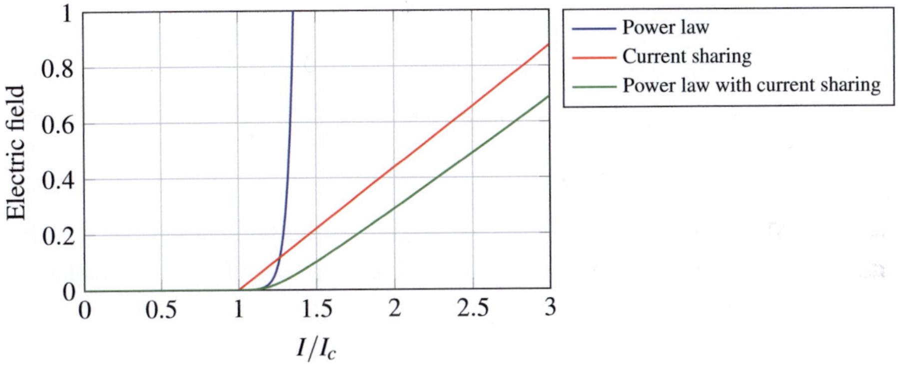

Figure 5.1: Current dependency of the electric field according the current sharing model (equation 5.1), the power law (equation 5.2), and the power law with current sharing (equation 5.3).

Ishiyama proposed a model where the current is shared between a power-law superconductor and a resistive stabilizer [134]. This relation is realistic both near the transition and at much higher currents. The total current I is shared between superconductor, which carries I_{sc}, and the resistive layer, which carries the remaining current. The electric field in superconductor and resistive layer are equal since they are connected in parallel:

$$E = E_c \left(\frac{I_{sc}}{I_c} \right)^n = \frac{\rho_e}{A}(I - I_{sc}) \tag{5.3}$$

This equation needs to be solved for I_{sc} before the electric field can be calculated. Because no exact solution exists, this must be done numerically. As seen in figure 5.1, the electric field matches the power law near the transition, while the slope equals the effective resistance of the current sharing model at higher currents.

5.2.2 Minimum propagating zones and quench energies

A local increase in temperature can create a normal zone where the current exceeds the critical current. The normal zone will either collapse or grow in time. A minimum propagating zone (MPZ) is the smallest normal zone that grows in time. Minimum propagating zones can be constructed by finding steady state solutions for the heat equation [109, p. 79]. The minimum quench energie (MQE) can be estimated by finding the lowest amount of energy required to form such a MPZ. For a one-dimensional wire in adiabatic conditions, one can find the following equation for the MQE:

$$\mathrm{MQE} = \frac{cA^2}{I}\sqrt{\frac{k(T_c - T_{cs})}{\rho}}\left(\pi + 2\sqrt{2}\right)(T_{cs} - T_0) \tag{5.4}$$

In this equation, c and k are the effective volumetric heat capacity and thermal conductivity, and T_{cs} is the current sharing temperature given by:

$$T_{cs} = T_0 + (T_c - T_0)\left(1 - \frac{I}{I_{c,0}}\right) \tag{5.5}$$

where $I_{c,0}$ is the critical current at the environmental temperature T_0. This method of finding the MQE is discussed further in appendix C. Unfortunately, it is of little use for coupled cables, because it does not consider current redistribution.

5.3 Electromagnetic and thermal model for partially coupled cables

A model was developed to simulate quench initiation in a Roebel cable, taking into account current redistribution between strands. The model comprises an electromagnetic and a thermal part. The thermal part calculates the temperature change due to thermal conduction and Joule heating. The electromagnetic part calculates the current redistribution in response to the temperature change. The thermal and electromagnetic systems are coupled and influence each other.

For simplicity it is assumed that the current density and temperature are homogeneous over the strand cross-section. Under this assumption, the state of an N-strand cable is described by N functions for the current $I_1(x,t), I_2(x,t), \ldots, I_N(x,t)$ and N functions for the temperature $T_1(x,t), T_2(x,t), \ldots, T_N(x,t)$. These unknown functions depend on the time t and the longitudinal coordinate x only. This simplification greatly reduces the complexity compared to a 3D model, while still providing useful insight into the dynamic behaviour of the cable. In this section, a coupled system of $2N$ PDEs for the unknown functions $I_i(x,t)$ and $T_i(x,t)$ will be derived. This system will then be solved numerically using a finite difference approximation.

5.3.1 Current diffusion equation for two parallel wires

Current redistribution between parallel wires can be described using diffusion-type PDEs [68, 117, 135]. Firstly, the diffusion equation will be derived for two parallel wires, and then it will be generalized for cables with any number of strands.

Consider two wires carrying a current $I_1(x,t)$ and $I_2(x,t)$. The sum of the current in the two wires is constant for any time and value of x:

$$I_{tot} = I_1(x,t) + I_2(x,t) \tag{5.6}$$

Let r be the longitudinal unit-length resistance of the wires in Ω/m, and ρ the unit-length resistance between the two wires in Ωm.[1] In case of superconducting wires, the longitudinal resistance r is a function of both current and temperature. The wires have a unit-length self-inductance M_{11}, M_{22} and are inductively coupled by a mutual inductance $M_{12} = M_{21}$ with units of H/m. Because of charge conservation, the derivative of the current with respect to x must equal the current arriving from other strands. Therefore, the transverse current per unit length flowing from strand 1 to strand 2 is:

$$\frac{\partial I_2}{\partial x} = -\frac{\partial I_1}{\partial x} \tag{5.7}$$

To find a differential equation for the current, the electric field integral along a loop ABCDA is considered (see figure 5.2). The loop forms a rectangle through the two wires and has a length a.

[1] In this chapter, the symbols R, r and ρ are used for quantities with units of Ω, Ω/m and Ω∗m respectively.

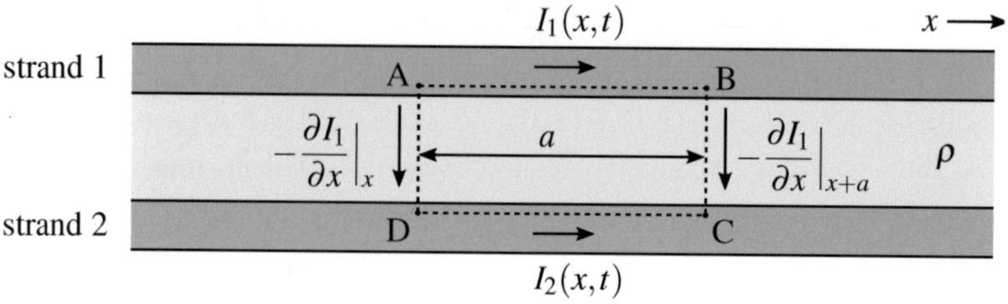

Figure 5.2: Two parallel wires with a transverse unit-length resistance ρ.

The electric field along paths AB and CD have resistive and inductive contributions, which in the first approximation depend only on the local current. The voltage over paths BC and DA equals the product of transverse current and resistance.

$$V_{AB} = a\left(I_1 r(I_1,T_1) + M_{11}\frac{\partial I_1}{\partial t} + M_{12}\frac{\partial I_2}{\partial t}\right) \tag{5.8}$$

$$V_{BC} = -\rho\frac{\partial I_1}{\partial x}\Big|_{x+a} \tag{5.9}$$

$$V_{CD} = -a\left(I_2 r(I_2,T_2) + M_{22}\frac{\partial I_2}{\partial t} + M_{12}\frac{\partial I_1}{\partial t}\right) \tag{5.10}$$

$$V_{DA} = \rho\frac{\partial I_1}{\partial x}\Big|_{x} \tag{5.11}$$

By equating the sum of these contributions to zero and using the fact that $\partial I_2/\partial x = -\partial I_1/\partial x$ one finds:

$$a\left((M_{11}+M_{22}-2M_{12})\frac{\partial I_1}{\partial t} + I_1 r(I_1,T_1) - I_2 r(I_2,T_2)\right) = \rho\frac{\partial I_1}{\partial x}\Big|_{x+a} - \rho\frac{\partial I_1}{\partial x}\Big|_{x} \tag{5.12}$$

Dividing by a and then taking the limit $a \to 0$ gives:

$$(M_{11}+M_{22}-2M_{12})\frac{\partial I_1}{\partial t} + I_1 r(I_1,T_1) - I_2 r(I_2,T_2) = \lim_{a\to 0}\frac{\rho\frac{\partial I_1}{\partial x}\big|_{x+a} - \rho\frac{\partial I_1}{\partial x}\big|_{x}}{a} = \frac{\partial}{\partial x}\rho\frac{\partial I_1}{\partial x} \tag{5.13}$$

The factor in front of the time derivative is the inductance of the return circuit of the two wires given by $L = M_{11}+M_{22}-2M_{12}$. Using the return circuit inductance the differential equation can be written as follows:

$$L\frac{\partial I_1}{\partial t} = \frac{\partial}{\partial x}\rho\frac{\partial I_1}{\partial x} + I_2 r(I_2,T) - I_1 r(I_1,T) \tag{5.14}$$

This is a diffusion equation for the current. The diffusion equation is widely studied and numerous analytic and numerical methods exist to solve it.

5.3.2 Current diffusion equation for an N-strand cable

In section 5.3.1, a current diffusion equation was derived for two parallel wires. It will now be generalized for cables with any number of strands. It is convenient to depict the inter-strand connections using a network of resistances. The network for two parallel wires is shown in figure 5.3. It can be seen as a schematic cross-section of the wires at a specific value of x. From the network, it is clear that the voltage between the two wires is $\rho \partial I_1/\partial x$.

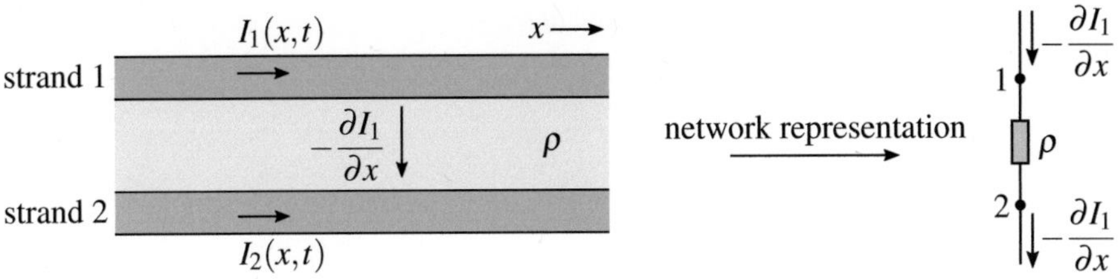

Figure 5.3: Network representation of two parallel wires with a transverse unit-length resistance ρ.

As shown in chapter 3, the inter-strand resistances in a Roebel cable can be described using a network model of two parameters: ρ_a and ρ_c. The parameter ρ_a is the unit-length resistance connecting adjacent strands and ρ_c is the 'crossing' resistance which connects non-adjacent as well. Both parameters have units of Ωm. The inter-strand resistance networks for Roebel cables with four and six strands are shown in figure 5.4. At each node numbered $i = 1,2,\ldots,N$, a current per unit length $-\partial I_i/\partial x$ flows into the network. Since no charge is accumulated in the network, the total current flow into the network must be zero:

$$\sum_{j=1}^{N} \frac{\partial I_j}{\partial x} = 0 \tag{5.15}$$

From this follows the spatial derivative of the current in strand N:

$$\frac{\partial I_N}{\partial x} = -\sum_{j=1}^{N-1} \frac{\partial I_j}{\partial x} \tag{5.16}$$

The relation between inter-strand currents and voltages can be found using Kirchhoff nodal analysis. Using the fact that the current into each node is zero, a system of $N-1$ linear equation can be constructed (see section 3.5 for a step-by-step derivation). The system in matrix form for any number of strands $N \geq 3$ is as follows:

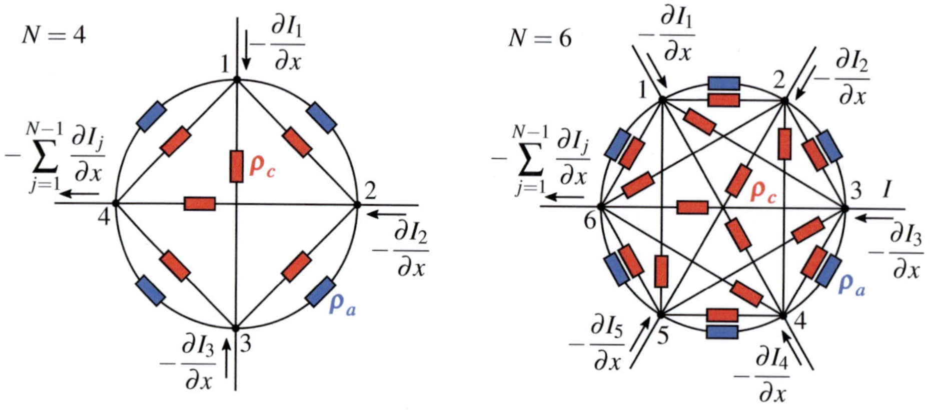

Figure 5.4: Inter-strand resistance network for Roebel cables with four and six strands.

$$\sigma \begin{bmatrix} U_1 - U_N \\ U_2 - U_N \\ U_3 - U_N \\ \vdots \\ U_{N-1} - U_N \end{bmatrix} = \frac{\partial}{\partial x} \begin{bmatrix} I_1 \\ I_2 \\ I_3 \\ \vdots \\ I_{N-1} \end{bmatrix} \tag{5.17}$$

with inter-strand conductance matrix σ given by:

$$\sigma = \begin{bmatrix} -\left(\frac{2}{\rho_a}+\frac{N-1}{\rho_c}\right) & \left(\frac{1}{\rho_a}+\frac{1}{\rho_c}\right) & \frac{1}{\rho_c} & \cdots & \frac{1}{\rho_c} \\ \left(\frac{1}{\rho_a}+\frac{1}{\rho_c}\right) & -\left(\frac{2}{\rho_a}+\frac{N-1}{\rho_c}\right) & \left(\frac{1}{\rho_a}+\frac{1}{\rho_c}\right) & \ddots & \vdots \\ \frac{1}{\rho_c} & \left(\frac{1}{\rho_a}+\frac{1}{\rho_c}\right) & -\left(\frac{2}{\rho_a}+\frac{N-1}{\rho_c}\right) & \ddots & \frac{1}{\rho_c} \\ \vdots & \ddots & \ddots & \ddots & \left(\frac{1}{\rho_a}+\frac{1}{\rho_c}\right) \\ \frac{1}{\rho_c} & \cdots & \frac{1}{\rho_c} & \left(\frac{1}{\rho_a}+\frac{1}{\rho_c}\right) & -\left(\frac{2}{\rho_a}+\frac{N-1}{\rho_c}\right) \end{bmatrix} \tag{5.18}$$

The off-diagonal elements of the matrix contain unit-length conductances between the corresponding strand pairs, which is $(1/\rho_a + 1/\rho_c)$ for adjacent and $1/\rho_c$ for non-adjacent strand pairs. The diagonal entries contain the negative sum of all conductances connected to the node. The vector on the right hand side contains the current flowing out of the network at each node.

More concisely, the system is expressed using index notation:

$$\sum_{j=1}^{N-1} \sigma_{ij}\,(U_j - U_N) = \frac{\partial I_i}{\partial x} \tag{5.19}$$

where σ_{ij} are the elements of the inter-strand conductance matrix:

$$\sigma_{ij} = \begin{cases} -\left(\frac{2}{\rho_a} + \frac{N-1}{\rho_c}\right) & i = j \\ \frac{1}{\rho_a} + \frac{1}{\rho_c} & |i-j| = 1 \\ \frac{1}{\rho_c} & |i-j| > 1 \end{cases} \tag{5.20}$$

In the next steps the aim is to derive a system of PDEs with $I_1, I_2, \ldots, I_N$ as only unknowns. To do this, $U_1, U_2, \ldots, U_N$ must be eliminated from equation 5.19. First, the matrix multiplication is inverted:

$$U_i - U_N = \sum_{j=1}^{N-1} \sigma_{ij}^{-1} \frac{\partial I_j}{\partial x} \tag{5.21}$$

In this equation, σ_{ij}^{-1} are the elements of the inverse matrix σ^{-1}. Differentiating with respect to x and noting that $\partial U/\partial x = -E$ one finds:

$$E_i - E_N = -\sum_{j=1}^{N-1} \frac{\partial}{\partial x} \sigma_{ij}^{-1} \frac{\partial I_j}{\partial x} \tag{5.22}$$

The electric field E has inductive and resistive contributions. In the first approximation, the inductive contribution can be expressed in the time derivative of local currents and mutual inductance M_{ij}. The resistive part is proportional to the non-linear longitudinal resistance r, which is a function of both current and temperature:

$$E_i = \underbrace{\sum_{j=1}^{N} M_{ij} \frac{\partial I_j}{\partial t}}_{\text{inductive}} + \underbrace{I_i r(I_i, T_i)}_{\text{resistive}} \tag{5.23}$$

By substituting this equation into 5.22, the electric field can be eliminated. This results in a system of diffusion-type equation with the currents $I_1, I_2, \ldots I_N$ as unknowns:

$$\sum_{j=1}^{N} (M_{ij} - M_{Nj}) \frac{\partial I_j}{\partial t} + I_i r(I_i, T_i) - I_N r(I_N, T_N) = -\sum_{j=1}^{N-1} \frac{\partial}{\partial x} \sigma_{ij}^{-1} \frac{\partial I_j}{\partial x} \tag{5.24}$$

A difficulty is that the left-hand side has a derivative of N unknown functions while the right hand side has only $N-1$. This can be overcome by assuming the total current is constant:

$$\sum_{j=1}^{N} \frac{\partial I_j}{\partial t} = 0 \quad \Rightarrow \quad \frac{\partial I_N}{\partial t} = -\sum_{j=1}^{N-1} \frac{\partial I_j}{\partial t} \tag{5.25}$$

The first term of equation 5.24 can now be rewritten as follows:

$$\sum_{j=1}^{N}(M_{ij}-M_{Nj})\frac{\partial I_j}{\partial t}=\sum_{j=1}^{N-1}(M_{ij}-M_{Nj})\frac{\partial I_j}{\partial t}+(M_{iN}-M_{NN})\frac{\partial I_N}{\partial t} \tag{5.26}$$

$$=\sum_{j=1}^{N-1}(M_{ij}-M_{Nj}-M_{iN}+M_{NN})\frac{\partial I_j}{\partial t} \tag{5.27}$$

It is convenient to express the inductance in terms of the return circuit inductance $L_{ij}=M_{ii}+M_{jj}-2M_{ij}$, because those are more easily calculated or found in literature:

$$M_{ij}-M_{Nj}-M_{iN}+M_{NN}=-\frac{1}{2}(M_{ii}+M_{jj}-2M_{ij})+\frac{1}{2}(M_{NN}+M_{jj}-2M_{Nj})$$

$$+\frac{1}{2}(M_{ii}+M_{NN}-2M_{iN}) \tag{5.28}$$

$$=-\frac{1}{2}(L_{ij}-L_{Nj}-L_{iN}) \tag{5.29}$$

For conciseness a $(N-1)\times(N-1)$ matrix Λ with the following elements is defined:

$$\Lambda_{ij}=-\frac{1}{2}(L_{ij}-L_{Nj}-L_{iN}) \tag{5.30}$$

The diffusion equation for the current can finally be written as:

$$\underbrace{\sum_{j=1}^{N-1}\Lambda_{ij}\frac{\partial I_j}{\partial t}}_{\text{inductive coupling}}=\underbrace{-\sum_{j=1}^{N-1}\frac{\partial}{\partial x}\sigma_{ij}^{-1}\frac{\partial I_j}{\partial x}}_{\text{conductive coupling}}\underbrace{+I_N r(I_N,T_N)-I_i r(I_i,T_i)}_{\text{longitudinal resistance}} \tag{5.31}$$

This is a system of $N-1$ independent equations. The terms in the equations all have units of V/m. The system becomes fully determined by considering that the total current does not change, and that the sum of the time derivatives of the strand currents is zero (equation 5.25):

$$\frac{\partial I_N}{\partial t}=-\sum_{j=1}^{N-1}\frac{\partial I_j}{\partial t} \tag{5.32}$$

For the special case $N=2$, the inductance matrix Λ_{11} equals L_{12} and the σ-matrix contains only a single value of $-1/\rho$. The diffusion equation then becomes:

$$L_{12}\frac{\partial I_1}{\partial t}=\frac{\partial}{\partial x}\rho\frac{\partial I_1}{\partial x}+I_2 r(I_2,T_2)-I_1 r(I_1,T_1) \tag{5.33}$$

This is consistent with the current diffusion equation for two wires which was derived in section 5.3.1.

5.3.3 Inductance of strand pairs

In order to evaluate the inductance matrix Λ (equation 5.30), it is necessary to find the return circuit inductance L_{ij} for each strand pair i and j. In a Roebel cable, each strand has the same geometry. Any pair of strands differs only by a longitudinal translation. Assuming an even spacing within a transposition length, strands i and j are longitudinally shifted by $|i-j|\ell_t/N$. The inductance of two strands picked from the cable depends on this longitudinal shift only. For example, strand pair 1-4 has the same inductance as pairs 2-5 and 3-6, because they are shifted by the same distance. The inductance thus has to be calculated only once for each relevant shifting distance.

The inductance of a return circuit can be found from the energy stored in the magnetic field. First, the magnetic field is calculated using the Biot-Savart law:

$$\mathbf{B}(x,y,z) = \frac{\mu_0}{4\pi} \iiint \frac{\mathbf{J} \times \mathbf{r}'}{|\mathbf{r}'|^3} \mathrm{d}V \tag{5.34}$$

where $\mathbf{r}'$ is a vector pointing from the source current at (x',y',z') to the observer at (x,y,z):

$$\mathbf{r}' = (x-x')\hat{\mathbf{x}} + (y-y')\hat{\mathbf{y}} + (z-z')\hat{\mathbf{z}} \tag{5.35}$$

The current is assumed to be homogeneously distributed over the width of the strands. Such assumption is necessary because in the model the current is a function of longitudinal coordinate x only. This is a drawback of the model. The consequences of this assumption will be further discussed in section 5.3.6.

The cable geometry repeats after every transposition length ℓ_t. Thus, the magnetic field is a periodic function of x with period ℓ_t. The average magnetic energy per unit length is therefore:

$$\varepsilon = \frac{1}{2\mu_0 \ell_t} \int_{-\infty}^{\infty} \int_{-\infty}^{\infty} \int_{0}^{\ell_t} |\mathbf{B}(x,y,z)|^2 \, \mathrm{d}x\, \mathrm{d}y\, \mathrm{d}z \tag{5.36}$$

Finally, the inductance is calculated using the formula for energy stored in an inductor:

$$\varepsilon = \frac{1}{2} L I^2 \quad \Rightarrow \quad L = \frac{2\varepsilon}{I^2} \tag{5.37}$$

Since the magnetic energy is proportional to the current squared, this formula yields a value independent of the current.

The integrals were evaluated numerically for the strand pairs of the cable investigated in this chapter. The cable parameters are listed in table 5.1. The inductance per unit length for any strand pair i,j are shown in table 5.2. The inductance increases with the longitudinal shift until a maximum is reached for half the transposition length.

Table 5.1: Cable parameters used for the inductance calculations.

	symbol	value	unit
Cable width	W	12.0	mm
Strand width	w_s, w_c	5.5	mm
Strand thickness	d	0.10	mm
Transposition length	ℓ_t	126	mm
No. of strands	N	10	-

Table 5.2: Return circuit inductance of strand pairs for the cable parameters in table 5.1.

$\lvert i-j \rvert$	shift [mm]	L_{ij} [10^{-7} H/m]
1	12.6	1.5397
2	25.2	2.9485
3	37.8	4.2277
4	50.4	5.4287
5	63.0	6.2848
6	75.6	5.4287
7	88.2	4.2277
8	100.8	2.9485
9	113.4	1.5397

5.3.4 Thermal equations

The temperature T_i in strand i is governed by the one-dimensional heat equation:

$$c(T_i)\frac{\partial T_i}{\partial t} = \frac{\partial}{\partial x} k_l(T_i)\frac{\partial T_i}{\partial x} + q_{trans,i} + q_{int,i} + q_{ext,i} \tag{5.38}$$

In this equation, $c(T_i)$ is the volumetric heat capacity in $\mathrm{JK^{-1}m^{-3}}$ and k_l is the longitudinal thermal conductivity in $\mathrm{WK^{-1}m^{-1}}$, both of which are functions of temperature (see section 5.4). The terms $q_{trans,i}$, $q_{int,i}$ and $q_{ext,i}$ represent contribution from transverse (strand-to-strand) heat transfer, internal and external heating. The dimension of the heat equation is power per unit volume and each term has units of $\mathrm{Wm^{-3}}$.

Transverse heat transfer in a Roebel cable is a complicated process governed by volumetric and surface thermal conductivities of the internal layers of each strand as well as the contact surface between strands. It is not feasible to calculate the temperature on such small scale for a long length cable. Instead, the effective transverse thermal conductivities of coated conductor stacks is used, for which experimental values are available from literature [136–138]. Let k_t

be the transverse thermal conductivity of a coated conductor stack. The temperature gradient between two strands within the stack is:

$$\frac{T_{i+1} - T_i}{d} \tag{5.39}$$

where d is the tape thickness. The heat transfer per unit length from strand i to $i+1$ is therefore:

$$-w_{con} k_t \frac{T_{i+1} - T_i}{d} \tag{5.40}$$

where w_{con} is the width of the contact surface. Average values for w_{con} for several Roebel cable architectures were calculated in section 3.2. To find the heat transverse per unit volume, the unit-length value is divided by the cross-sectional area of the strand A:

$$-w_{con} k_t \frac{T_{i+1} - T_i}{Ad} \tag{5.41}$$

In a Roebel cable, every strand i touches only its neighbours $i-1$ and $i+1$. The transverse heat flow towards strand i is therefore:

$$q_{trans,i} = k_t w_{con} \frac{T_{i+1} - T_i}{Ad} + k_t w_{con} \frac{T_{i-1} - T_i}{Ad} = \frac{k_t w_{con}}{Ad} \left(T_{i-1} - 2T_i + T_{i+1} \right) \tag{5.42}$$

Because of full transposition, strands 1 and N also form a strand pair with contact width w_{con}. The transverse heat flows to those strands are thus given by:

$$q_{trans,1} = \frac{k_t w_{con}}{Ad} \left(T_N - 2T_1 + T_2 \right) \tag{5.43}$$

$$q_{trans,N} = \frac{k_t w_{con}}{Ad} \left(T_{N-1} - 2T_N + T_1 \right) \tag{5.44}$$

These linear equations can be expressed in matrix form to find q_{trans} for every strand:

$$\begin{bmatrix} q_{trans,1} \\ q_{trans,2} \\ q_{trans,3} \\ \vdots \\ q_{trans,N} \end{bmatrix} = \frac{k_t w_{con}}{Ad} \begin{bmatrix} -2 & 1 & & & 1 \\ 1 & -2 & 1 & & \\ & \ddots & \ddots & \ddots & \\ & & 1 & -2 & 1 \\ 1 & & & 1 & -2 \end{bmatrix} \begin{bmatrix} T_1 \\ T_2 \\ T_3 \\ \vdots \\ T_N \end{bmatrix} \tag{5.45}$$

or, in short:

$$q_{trans,i} = \frac{k_t w_{con}}{Ad} \sum_{j=1}^{N} D_{ij} T_j \quad \text{where} \quad D_{ij} = \begin{cases} -2 & i = j \\ 1 & |i-j| = 1 \quad \text{or} \quad |i-j| = N-1 \\ 0 & \text{otherwise} \end{cases} \tag{5.46}$$

The internal Joule heating per unit length is $I_i^2 r(I_i, T_i)$ where $r(I_i, T_i)$ is the non-linear resistance per unit length. The internal heating per unit volume is therefore:

$$q_{int,i} = \frac{I_i^2 r(I_i, T_i)}{A} \tag{5.47}$$

Substituting the equations for transverse conduction and internal heating into equation 5.38, the thermal equation becomes:

$$\underbrace{c(T_i)\frac{\partial T_i}{\partial t}}_{\substack{\text{heat}\\\text{capacity}}} = \underbrace{\frac{\partial}{\partial x} k_l(T_i) \frac{\partial T_i}{\partial x}}_{\substack{\text{longitudinal}\\\text{conduction}}} + \underbrace{\frac{k_t w_{con}}{Ad} \sum_{j=1}^{N} D_{ij} T_j}_{\substack{\text{transverse}\\\text{conduction}}} + \underbrace{\frac{I_i^2 r(I_i, T_i)}{A}}_{\substack{\text{internal}\\\text{heating}}} + \underbrace{q_{ext}}_{\substack{\text{external}\\\text{heating}}} \tag{5.48}$$

5.3.5 Numerical solution using a finite difference approximation

In sections 5.3.2 and 5.3.4, a system of coupled PDEs describing electromagnetic and thermal behaviour of the Roebel cable was derived. The system consist of $2N$ equations and has $2N$ unknown functions, namely current functions $I_1(x,t), I_2(x,t), \ldots, I_N(x,t)$ and temperature functions $I_1(x,t), I_2(x,t), \ldots, I_N(x,t)$. The system is summarized below:

$$\text{Current diffusion:} \quad \underbrace{\sum_{j=1}^{N-1} \Lambda_{ij} \frac{\partial I_j}{\partial t}}_{\text{inductive coupling}} = \underbrace{-\sum_{j=1}^{N-1} \frac{\partial}{\partial x} \sigma_{ij}^{-1} \frac{\partial I_j}{\partial x}}_{\text{conductive coupling}} \underbrace{+ I_N r(I_N, T_N) - I_i r(I_i, T_i)}_{\text{longitudinal resistance}} \tag{5.49}$$

$$\text{Current sum:} \quad \sum_{j=1}^{N} \frac{\partial I_j}{\partial t} = 0 \tag{5.50}$$

$$\text{Heat equation:} \quad c(T_i)\frac{\partial T_i}{\partial t} = \underbrace{\frac{\partial}{\partial x} k_l(T_i) \frac{\partial T_i}{\partial x}}_{\substack{\text{longitudinal}\\\text{conduction}}} + \underbrace{\frac{k_t w_{con}}{Ad} \sum_{j=1}^{N} D_{ij} T_j}_{\substack{\text{transverse}\\\text{conduction}}} + \underbrace{\frac{I_i^2 r(I_i, T_i)}{A}}_{\substack{\text{internal}\\\text{heating}}} + \underbrace{q_{ext}}_{\substack{\text{external}\\\text{heating}}} \tag{5.51}$$

Equations 5.49 and 5.51 are one-dimensional diffusion equations. Such equations can be numerically solved using a finite difference approximation [139]. To do this, the space dimension x is discretized with equal spacing Δx:

$$x_p = p \cdot \Delta x, \quad p = 0, 1, 2, \ldots, P \tag{5.52}$$

This creates a grid of points at which the currents and temperatures are calculated. The grid is illustrated by figure 5.5.

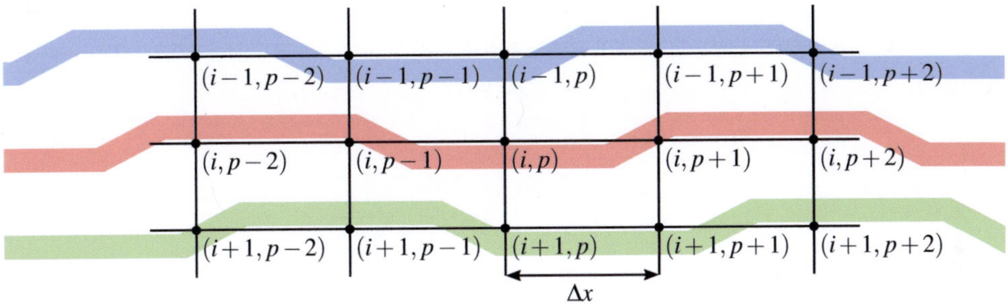

Figure 5.5: Grid for the finite-difference approximation.

The terms with derivatives in x are approximated using finite differences. Assuming that the inter-strand conductance matrix σ is independent of x, the conductive coupling term can be approximated as follows:

$$\frac{\partial}{\partial x}\sigma_{ij}^{-1}\frac{\partial I_j}{\partial x} = \sigma_{ij}^{-1}\frac{\partial^2 I_j}{\partial x^2} \approx \sigma_{ij}^{-1}\frac{I_{j,p+1} - 2I_{j,p} + I_{j,p-1}}{(\Delta x)^2} \tag{5.53}$$

with

$$I_{j,p} = I_j(x_p, t) \tag{5.54}$$

The longitudinal heat conductivity depends on temperature. Morton and Mayers recommend the following approximation for a derivative in this form [139, p. 51]:

$$\frac{\partial}{\partial x}k_l(T_i)\frac{\partial T_i}{\partial x} \approx \frac{k_l\left(T_{i,p+1/2}\right)\left(T_{i,p+1} - T_{i,p}\right) - k_l\left(T_{i,p-1/2}\right)\left(T_{i,p} - T_{i,p-1}\right)}{(\Delta x)^2} \tag{5.55}$$

where

$$T_{i,p} = T_i(x_p, t) \tag{5.56}$$

$$T_{i,p\pm 1/2} = \frac{T_i(x_p, t) + T_i(x_{p\pm 1}, t)}{2} \tag{5.57}$$

To evaluate the finite differences at $p = 0$ and $p = P$, boundary conditions are needed. In the quench simulations, the external disturbance is applied at $x = 0$. Both temperature and the current distribution will therefore be even functions of x, which leads to the following boundary condition:

$$I_{i,-1} = I_{i,1} \tag{5.58}$$

$$T_{i,-1} = T_{i,1} \tag{5.59}$$

Using this symmetry, only one half of the cable needs to be simulated, reducing computation time. A constant temperature T_0 is enforced on the cable ends, leading to a Dirichlet boundary condition:

$$T_{i,P} = T_0 \tag{5.60}$$

For the current, there are two possible ways to define boundary conditions. Either a fixed current distribution is enforced on the cable ends, or the current is allowed to redistribute freely. The first case describes a cable with very high resistance to the current leads, or a cable which is driven by separate current sources for each strand. Such a situation is described by Dirichlet boundary condition with a constant current I_0:

$$I_{i,P} = I_0 \tag{5.61}$$

In the second case, the current can redistribute freely between the strands at the cable ends. In other words, the strands are shot-circuited, and the transverse voltage at the ends is zero. By equation 5.19, transverse voltages are proportional to the x-derivatives of the current functions. In case of an electrical short-circuit, the x-derivatives therefore equals zero at the cable ends. This leads to homogeneous Neumann boundary conditions:

$$I_{i,P+1} = I_{i,P} \tag{5.62}$$

To simulate long-length cables, the discrete cable length should be so long that the altered current distribution at $x = 0$ never reaches the end of the cable. In this case, the boundary conditions at $x = P\Delta x$ do not influence the solution.

By using finite difference approximations for the derivatives in x, only derivatives in time remain. The system of PDEs therefore becomes a system of ODEs. The ODEs are numerically integrated using Matlab's build-in ODE solver 'ode15s'. This is a solver with a variable time step that is automatically adjusted to meet an error criterion. A relative error tolerance of 10^{-5} was used for all calculations in this chapter.

5.3.6 Limitations of the calculation method

Current redistribution can occur over a length much longer than the normal zone itself. For an accurate result, the simulated cable should be at least as long as the region in which current transfer occurs. The calculation therefore should have a reasonable performance even for long cables. By using one-dimensional equations for each strand, the number of unknowns is kept as low as possible. The model is the simplest representation of the cable that still takes into account current redistribution between the strands. The simplifications, however, also have consequences for the accuracy of the calculation:

- The current within each strand is assumed to be evenly distributed, in order to calculate the inductance of strand pairs. In reality, a transport current fills the strands from the outside with the critical current density [133, p. 162]. This distribution will change in response to a temperature rise. Inductive coupling between strands is therefore more complicated than assumed. Also, induced screening currents and inter-layer resistances are ignored.
- The diffusion equation used to calculate the current distribution considers inductive coupling only locally. This means that the induced electric field at x depends only on the time derivative of the current distribution at the same value of x. In case of sharp changes in current density, this will produce a deviation.
- The temperature is assumed to be a function of the longitudinal coordinate x only. Temperature gradients in the lateral direction or between the layers within the strands are ignored.
- The three-dimensional geometry of the Roebel cable is not considered. The matrices Λ, σ and D only describe length-averaged inductive, conductive and thermal coupling. The consequence is that disturbances to specific locations through which the strands transpose cannot be simulated (e.g. the cross-over to the other cable half).
- Heat transfer to the environment is ignored, except at the cable ends when Dirichlet boundary conditions are used.

5.4 Temperature-dependent properties

There are large differences in temperature during a quench. Besides the critical current, also the specific heat, electrical resistivity and thermal conductivity depend on temperature. The temperature dependence of these physical properties needs to be taken in to account for an accurate simulation. This section describes these temperature-dependent properties for SuperPower SCS12050-AP coated conductor tape, which was used in all quench measurements.

This tape is 12 mm wide, has a 50 μm thick Hastelloy C-276 substrate and is coated with 20 μm of copper surrounding the tape.

5.4.1 Critical current

The critical current of a short sample was measured at temperatures ranging from 61 to 92 K on a sample holder cooled by helium gas. The measurements were done in magnetic self-field. The voltage was measured over a distance of 3 cm, and a electric field criterion of $E_c = 10^{-4}$ V/m was used to determine the critical current. The set-up used for these measurements is described in more detail in a work by Liu et al. [140]. The critical current was also measured in in a liquid nitrogen bath with a temperature of 77.3 K. The critical current at this temperature was 346 A. The measurement data are shown in figure 5.6.

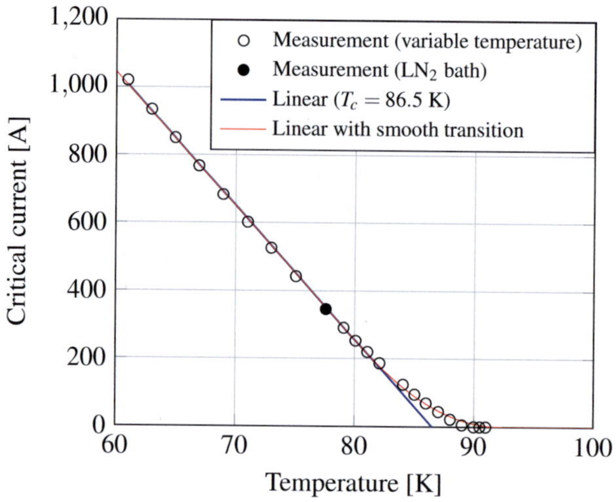

Figure 5.6: Critical current as a function of temperature for SuperPower SCS12050-AP coated conductor with linear (equation 5.63) and smooth (equation 5.65) fits. All measurements were done by Mayraluna Lao.

The simplest empirical $I_c(T)$ relation decreases linearly with temperature and reaches at zero at the some critical temperature T_c. This model is used in analytical calculations of stability, including the minimum propagating zone method. Using the critical current measured at 77.3 K as a reference point, the linear $I_c(T)$ model is:

$$I_{c,linear}(T) = \begin{cases} I_{c,77.3\text{ K}} \left(1 - \dfrac{T - 77.3\text{ K}}{T_c - 77.3\text{ K}}\right) & T \leq T_c \\ 0 & T \geq T_c \end{cases} \tag{5.63}$$

A reasonable fit can be found for temperatures below 83 K using a critical temperature of $T_c = 86.5$ K, as seen in figure 5.6. However, this fit underestimates the critical current near the transition, because the transition is not as abrupt as the linear model assumes.

In order to improve accuracy near the transition, the linear model is smoothed by calculating a moving average over a temperature range $T \pm \delta$. This results in a parabolic $I_c(T)$ relation for temperatures between $T_c - \delta$ and $T_c + \delta$:

$$I_{c,smooth}(T) = \frac{1}{2\delta} \int_{T-\delta}^{T+\delta} I_{c,linear}(T')\,\mathrm{d}T' \tag{5.64}$$

$$= \begin{cases} I_{c,linear}(T) & T \leq T_c - \delta \\ I_{c,77.3\ \mathrm{K}} \dfrac{(T_c - T + \delta)T_c - \frac{1}{2}(T_c^2 - (T-\delta)^2)}{2\delta(T_c - 77.3\ \mathrm{K})} & T_c - \delta \leq T \leq T_c + \delta \\ 0 & T \geq T_c + \delta \end{cases} \tag{5.65}$$

Using least-squares fitting an optimal value of $\delta = 5.7$ K was found. This formula will be used for all quench simulations in this chapter. To find the $I_c(T)$ of a Roebel cable, only self-field critical current in liquid nitrogen $I_{c,77.3\ \mathrm{K}}$ is changed. It is thus assumed that the self-field reduction of the cable critical current is independent of temperature. In this way, it is not necessary to do time consuming temperature-dependent I_c measurements on Roebel cables.

5.4.2 Resistance of the stabilizing layers

If the critical current is exceeded, the applied current is shared between the superconductor and metal layers in parallel. The electrical resistivity of the stabilizer is needed in order to calculate the electric field using one of the current sharing models (section 5.2.1).

The electrical resistivity ρ of a metal can be expressed as the sum of an ideal resistivity ρ_i and residual resistivity ρ_{res}:

$$\rho(T) = \rho_i(T) + \rho_{res} \tag{5.66}$$

The ideal resistivity ρ_i results from phonon scattering and is temperature dependent. Tabulated values for ρ_i over a wide range of temperatures are available in a work from Ekin [120, p. 575]. The residual resistivity ρ_{res} represents the contribution from defects and is usually expressed in the "residual resistivity ratio" defined as:

$$\mathrm{RRR} = \frac{\rho_{295\ \mathrm{K}}}{\rho_{res}} \approx \frac{\rho_{295\ \mathrm{K}}}{\rho_{4\ \mathrm{K}}} \tag{5.67}$$

Residual resistive ratio (RRR) values were determined by Bonura et al. for copper extracted from different coated conductors [141]. Depending on the manufacturer, the RRR-value was found to range from 17 to 61 (see table 5.3). The ideal resistivity of copper as a function of temperature and total resistivity for different RRR values are shown in figure 5.7.

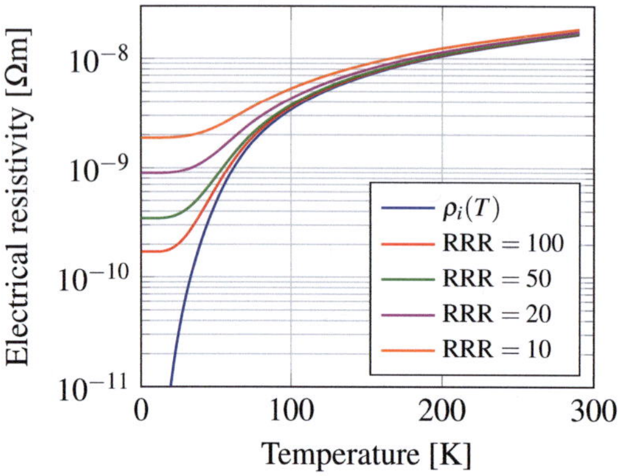

Figure 5.7: Ideal resistivity of copper $\rho_i(T)$ (data from Ekin [120, p. 575]) and resistivity for copper with different residual resistivity ratios (RRR) computed using equations 5.66 and 5.67.

Table 5.3: RRR values of copper extracted from different coated conductors measured by Bonura et al. [141].

Manufacturer	RRR
AMSC	19
Bruker HTS	17
Fujikura	59
SuNAM	61
SuperOx	14
SuperPower	42

At cryogenic temperatures, the electrical resistivity of the substrate material Hastelloy C-276 is $123 * 10^{-8}$ Ωm with only a weak temperature dependency [142]. For different types of stainless steel it is in the range 49-73 $*10^{-8}$ Ωm [120, p. 576]. These values are at least two orders of magnitude higher than the resistivity of copper. In copper-stabilized tapes, the substrate carries only a small fraction of the current, even during a quench.

The resistance of the coated conductor sample described in section 5.4.1 was measured at 94 and 100 K. These temperatures exceed the critical temperature, leading to a linear I-V-curve of which the slope equals the resistance. The measured resistance values are shown in figure 5.8. The calculated resistance of the copper layer assuming a thickness $d_{cu} = 40$ µm and a RRR of 42 is also shown in the figure. Since silver and Hastelloy layers are in parallel with the copper layers, one expects the measured resistance to be slightly lower than the theoretical value. However, as seen in figure 5.8, the measured resistance exceeds that of a 40 µm copper layer. This means that either the copper layer is thinner than listed in the datasheet or its RRR is lower. A better match is found using a copper thickness of 30.8 µm. This value will be used in the calculations from now on.

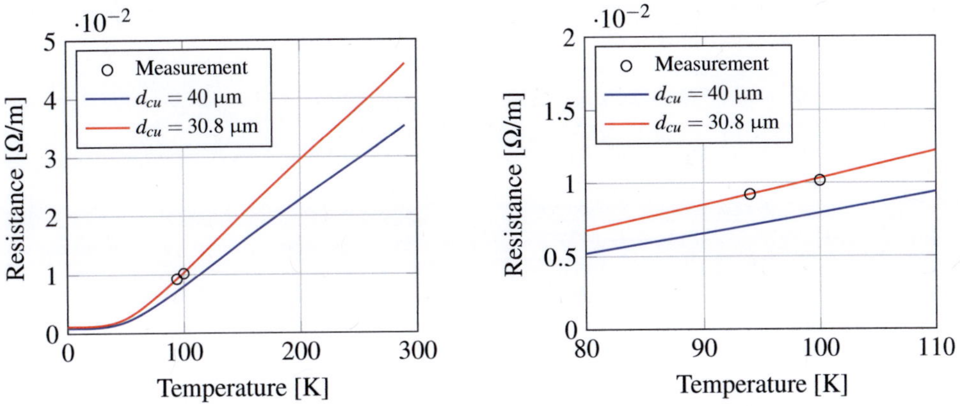

Figure 5.8: Measured resistance per unit length at 94 and 100 K and resistance calculated from the copper resistivity assuming RRR = 42 and a copper thickness d_{cu}. The match with measured values is better when a copper thickness $d_{cu} = 30.8$ µm is used instead of the datasheet value $d_{cu} = 40$ µm. The right figure shows a close-up near the two measurement points.

5.4.3 Thermal conductivity and heat capacity

The thermal conductivity of copper and the common substrate materials stainless steel and Hastelloy C-276 are shown in figure 5.9. In the cryogenic temperature range, the thermal conductivity of copper is two to four orders of magnitude higher than that of the substrate materials. In copper-stabilized tapes, the copper is therefore expected to dominate thermal conductance.

The thermal conductivity in coated conductor tapes was measured at various temperatures by Bagrets et al. [143], and is also shown in figure 5.9. Tapes from SuperPower with and without copper were measured. The copper stabilization was found to dominate the thermal conductivity. These data are used in the calculation, since the same type of coated conductor is used in the investigated Roebel cables.

The volumetric heat capacity of the coated conductor is needed to calculate the rate at which the temperature changes. The heat capacity of metals above about 10 K is dominated by the phonon contribution and can be estimated using Debye's formula:

$$c(T) = 9Nk_B \left(\frac{T}{T_D}\right)^3 \int_0^{T/T_D} \frac{x^4 e^x}{(e^x-1)^2} \mathrm{d}x \tag{5.68}$$

Here N is the number of atoms per unit volume, k_B is Boltzmann's constant and T_D the material-specific Debye temperature. The Debye temperature is 343.5 K [144] for copper and 373 K for Hastelloy C-276 [142].

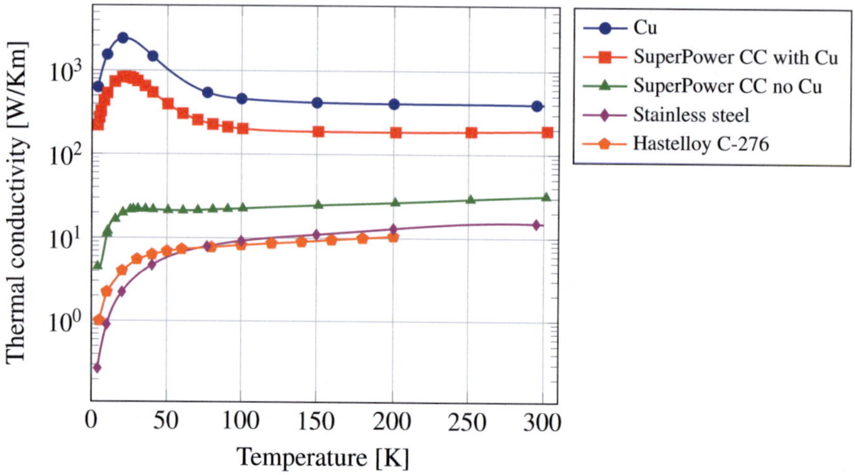

Figure 5.9: Measurements of longitudinal thermal conductivity in a SuperPower coated conductor by Bagrets et al. [143]. The thermal conductivity of copper, stainless steel [120, p. 579] and Hastelloy C-276 [142] are shown for comparison.

To find the volumetric heat capacity of the coated conductor tape, a weighted average of the values for copper and Hastelloy C-276 is used:

$$c_{coated\ conductor}(T) = \frac{d_{cu}c_{cu}(T) + d_{has}c_{has}(T)}{d_{cu} + d_{has}} \tag{5.69}$$

The effect of the much thinner *RE*BCO, silver and buffer layers on the volumetric heat capacity is neglected. The volumetric heat capacity for both materials and the weighted average as a function of temperature are shown in figure 5.10.

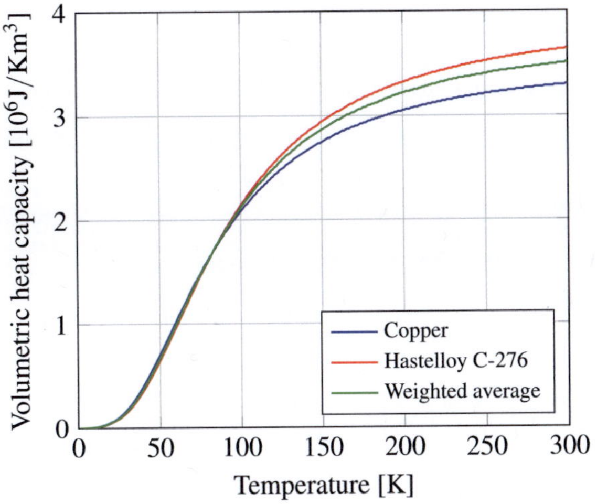

Figure 5.10: Volumetric heat capacities of copper and Hastelloy C-276 as a function of temperature calculated using Debye's formula. The volumetric heat capacity of the coated conductor is estimated by taking a weighted average with $d_{has} = 50$ µm and $d_{cu} = 30.8$ µm.

5.5 Simulations of externally induced quenches

5.5.1 Single coated conductor

In order to validate the thermal model, the predicted minimum quench energies for single tapes are compared to a study by Pelegrín et al. [145]. Pelegrín measured the minimum quench energy of a 4-mm-wide coated conductor from SuperPower in adiabatic conditions at 72 and 77 K. The sample had a 50 µm thick substrate of Hastelloy and a 1 µm thick layer of *RE*BCO. It was stabilized with 2 µm of silver deposited on the *RE*BCO layer and 20 µm of electroplated copper surrounding the tape. The critical current of the sample was 88 A at 77 K. A 2-mm-long graphite heater was placed on the middle of the sample. To find the minimum quench energy, 40 ms long heat pulses of different magnitude were applied while the tape carried a constant current. The current was shut down 0.5 to 1 s after the heat pulse, in order to prevent damage to the sample.

The measured MQE values ($T = 77$ K) and a simulation by Pelegrín et al. are shown in figure 5.11. The measurement values exceed the calculated ones by about 30%. Pelegrín attributed this difference to the increased heat capacity of the sample near the heater. Pelegrín's results are compared to three different MQE predictions using the theory of this chapter: Numerical calculations using the current sharing (equation 5.1) and the power-law current sharing (equation 5.3) descriptions of the electric field are shown. Also, MQE calculated using the method of minimum propagating zones are shown, which was described in section 5.2.2. The simulation using the power law current sharing method matches the simulation by Pelegrín. This confirms the validity of the thermal model. This method also makes the best match with the experimental data, and therefore it will be used for the Roebel cables MQE calculations in the next section. Interestingly, the minimum propagating zone method provides a good estimate without the need for numerical calculations. This method, however, only works for constant current distributions, and cannot be used for Roebel cables with coupled strands.

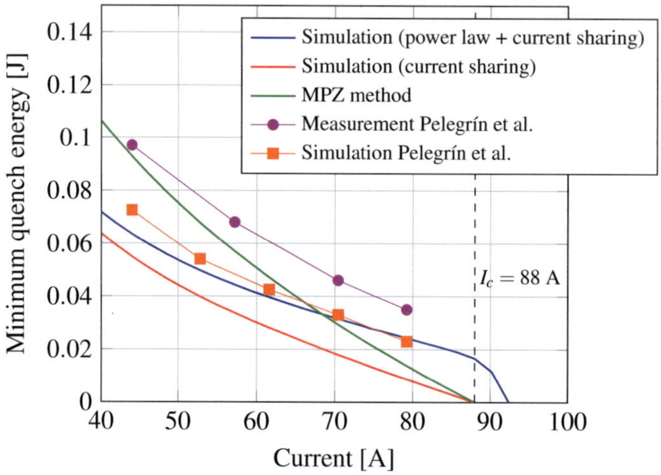

Figure 5.11: Minimum quench energies of a single tape at $T = 77$ K.

5.5.2 Roebel cables with different levels of coupling

In order to explore the effect of inter-strand resistance and transverse thermal conductivity on the minimum quench energy, a large number of MQE calculations were done for different values of those parameters. The simulated cable has 10 strands with a critical current per strand of $I_c(77.3\text{ K}) = 106$ A, equal to that of the cable used in the MQE measurements (see section 5.6). The adjacent transverse inter-strand resistance ρ_a is ranged from 10^{-7} Ωm to infinity. An adjacent inter-strand resistance of 10^{-7} Ωm is on the lower limit for a solder-filled cable, while an infinite inter-strand resistance represents a cable with insulated strands. The crossing inter-strand resistance is infinite, allowing current redistribution between neighbouring strands only. The transverse thermal conductivity k_t is ranged from 0.001 to 1 W/Km. This covers the ranges of transverse thermal conductivities measured in coated conductor stacks [136, 137]. All temperature-dependent properties are as described in section 5.4. The simulated cable has a length of 10 m, with a fixed current distribution on both ends. Current redistribution is thus only possible over this 10 m length. For cables longer than 10 m, this is a pessimistic approach, because current redistribution can occur over an even longer length. The 10 m length was chosen for performance reasons. A heat pulse is applied to the middle of strand 1 for a time of 0.1 s. A quench is recorded if any of the strands develops a voltage exceeding 10 mV within 10 seconds. This criterion is the same as the one used in the experiment. The technical reasons for this criterion will be discussed in the next section.

In figures 5.12-5.15, the calculated MQE values are shown, all of which were done for a starting temperature of $T_0 = 77.0$ K. In each figure, the transverse thermal conductivity k_t is fixed, while the different values of the inter-strand resistance ρ_a are represented by different lines. The MQEs of a single strand are shown for comparison. The effect of inter-strand resistance is the largest if the transverse thermal conductivity is low (figure 5.12). For insulated strands ($\rho_a = \infty$ Ωm), the MQE is close to that of a single strand: a quench in one strand causes a quench in the entire cable, because the current cannot redistribute. For low inter-strand resistance ($\rho_a = 10^{-7}$ Ωm), the MQE is much higher, and even exceeds that of the single strand value multiplied by 10. The low thermal conductivity keeps the hotspot confined to strand 1, while the low inter-strand resistance allows the current to redistribute to other strands. This prevents additional Joule heating in the hotspot, and allows the cable to recover.

For higher values of the transverse thermal conductivity (figures 5.13-5.15), the effect of inter-strand resistance becomes smaller. If the transverse thermal conductivity is 1 W/Km, the MQE approaches the sum of single strand MQE, regardless of the inter-strand resistance. In this case, the heat pulse is distributed over all the strands very quickly, so that the cable behaves much like a single large strand.

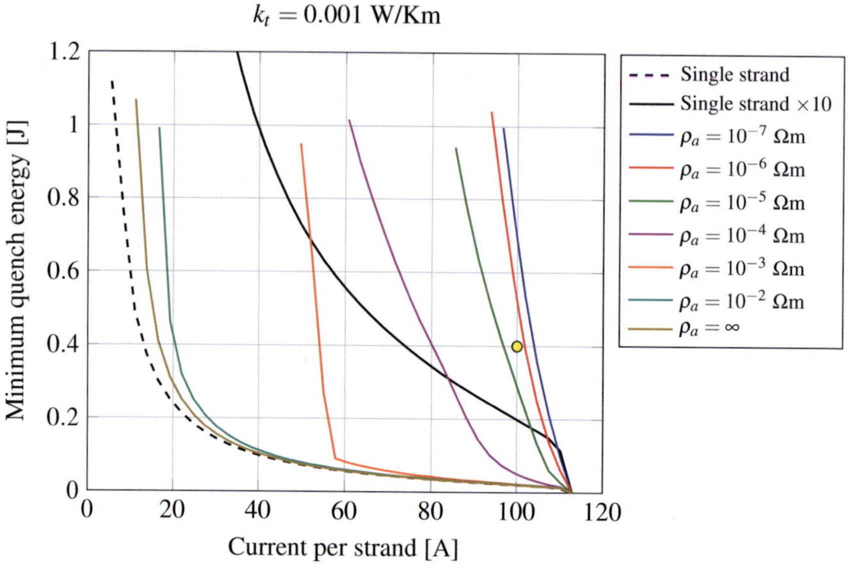

Figure 5.12: Calculated minimum quench energy of a Roebel cable with transverse thermal conductivity $k_t = 0.001$ W/Km and different values of the adjacent inter-strand resistance ρ_a.

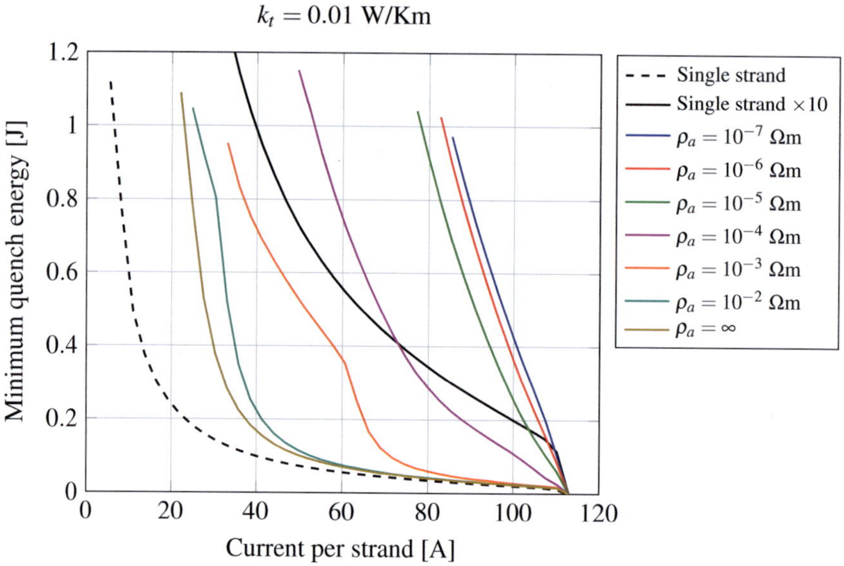

Figure 5.13: Calculated minimum quench energy of a Roebel cable with transverse thermal conductivity $k_t = 0.01$ W/Km and different values of the adjacent inter-strand resistance ρ_a.

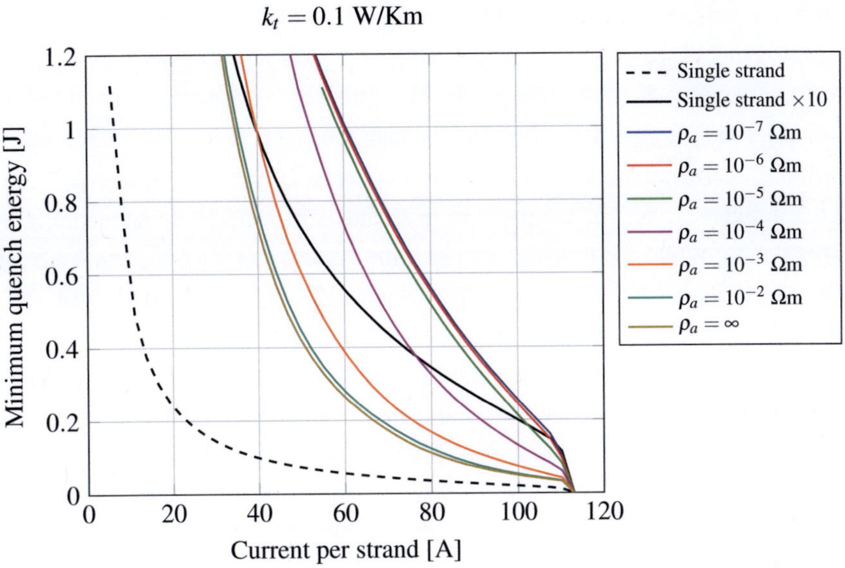

Figure 5.14: Calculated minimum quench energy of a Roebel cable with transverse thermal conductivity $k_t = 0.1$ W/Km and different values of the adjacent inter-strand resistance ρ_a.

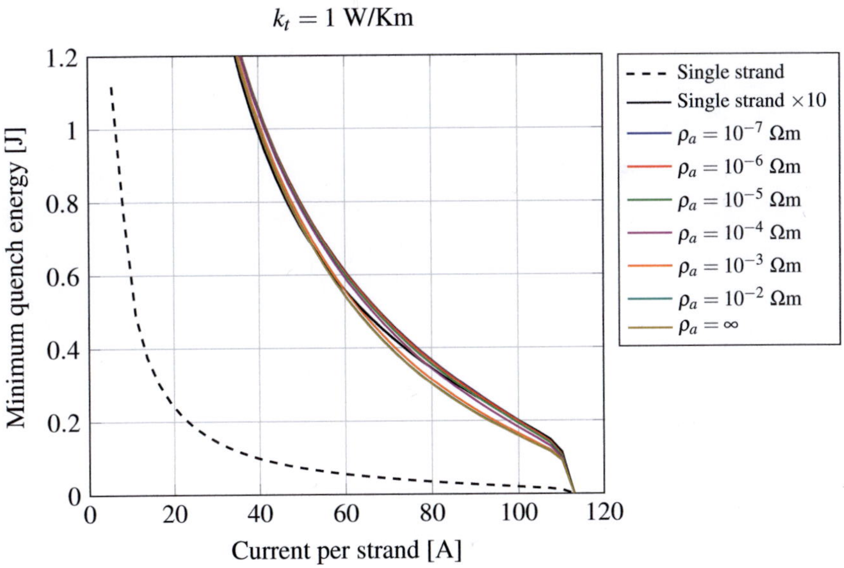

Figure 5.15: Calculated minimum quench energy of a Roebel cable with transverse thermal conductivity $k_t = 1$ W/Km and different values of the adjacent inter-strand resistance ρ_a.

To better understand the effect of current redistribution, two simulations that differ only in the inter-strand resistance will be compared. In figure 5.12, a yellow circle is shown at a current of 100 A per strand and a 0.4 J heat pulse. This situation leads to a quench if the adjacent inter-strand resistance ρ_a is 10^{-5} Ωm, but the cable recovers if it is 10^{-6} Ωm. The current distribution and temperature at the location of the heater ($x = 0$) for the case $\rho_a = 10^{-5}$ Ωm are shown in figure 5.16. The temperature in strand 1 rises to 154 K at the end of the heat pulse ($t = 0.1$ s). While the temperature rises, the current in strand 1 drops from 100 A to 1.7 A, greatly reducing Joule heating in the hotspot. However, as the heat pulse diffuses to other strands, the temperature rises there as well. At $t = 2.79$ s, superconductivity is lost in all strands. The temperature starts to rise quickly, because the current can no longer flow around the hotspot. Note that in this stage each strands carries approximately the same current of 100 A. At $t = 3.29$, a temperature of 273 K is reached and the simulation is stopped.

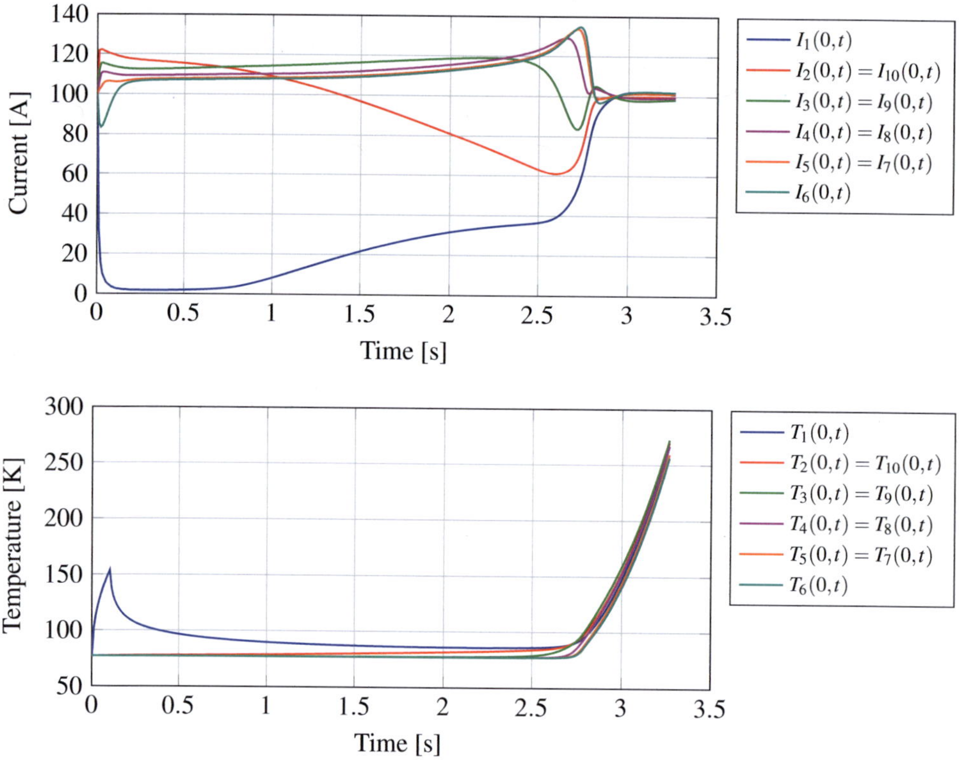

Figure 5.16: Calculated current distribution and temperatures at the location of the heater ($x = 0$). The cable reaches room temperature after 3.26 seconds. Simulation parameters: $\rho_a = 10^{-5}$ Ωm, $k_t = 0.001$ W/Km, $\varepsilon = 0.4$ J.

In the simulation shown in figure 5.17, the adjacent inter-strand resistance ρ_a was decreased by one order of magnitude to 10^{-6} Ωm. This makes current redistribution more effective. The current in strand 1 now drops to only 0.4 A. This reduces internal heating and prevents the quench from happening. As seen in figure 5.18, the internal heating is reduced by about 50%.

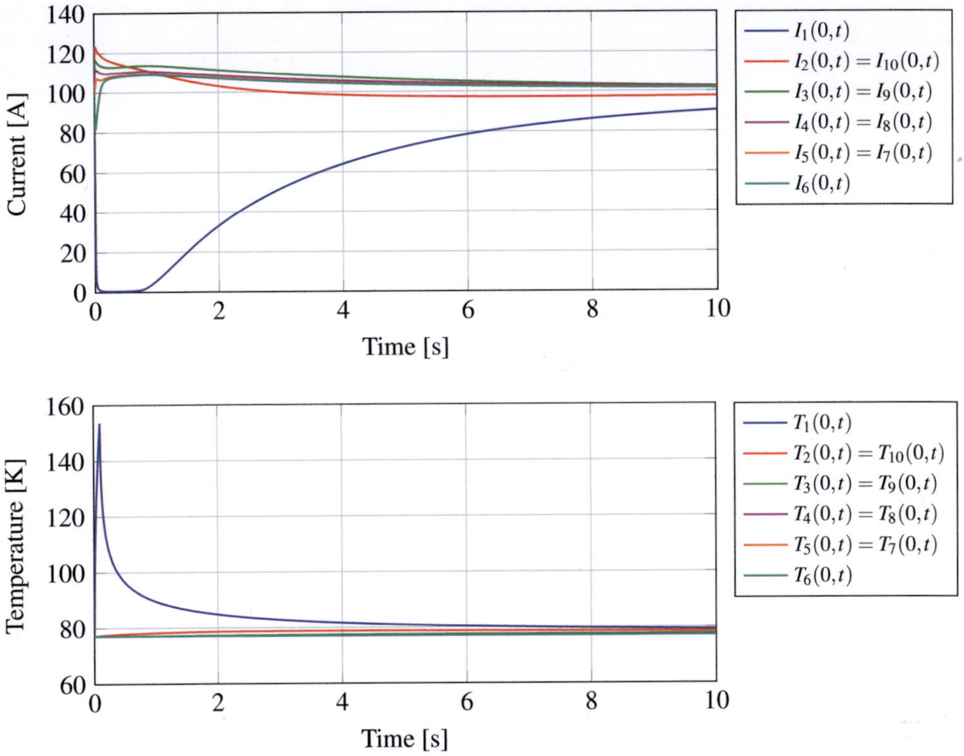

Figure 5.17: Calculated current distribution and temperatures at the location of the heater ($x = 0$). The cable has not quenched after 10 seconds. Simulation parameters: $\rho_a = 10^{-6}$ Ωm, $k_t = 0.001$ W/Km, $\varepsilon = 0.4$ J.

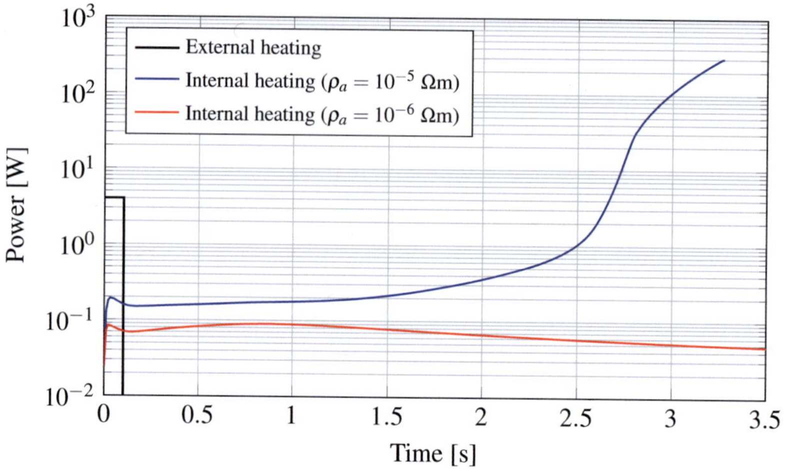

Figure 5.18: Total internal Joule heating after a 0.4 J heat pulse. Due to more effective current redistribution, the internal heating is lower for low values of ρ_a. In this specific case, a value of $\rho_a = 10^{-6}$ Ωm allows the cable to recover, while $\rho_a = 10^{-5}$ Ωm leads to a quench. The power is shown on a logarithmic axis to highlight the difference.

5.6 Minimum quench energy measurements

This section describes minimum quench energy measurements done on a Roebel cable. The minimum quench energy is determined by applying heat pulses of different magnitudes to one of the strands using a heater. The experiment is done in quasi-adiabatic conditions using a conduction-cooled sample holder in a vacuum chamber.

5.6.1 Experimental set-up

The minimum quench energy measurements were carried out in the Vatesta facility at ITEP. This facility comprises a 5 T solenoid magnet with a 800 mm warm bore and a variable-temperature insert cryostat. The magnet however was not used and all measurements were done in self-field.

A picture of the variable-temperature insert can be seen in figure 5.19. The principal part is a large mechanical support structure which transfers the Lorentz forces to the lid of the cryostat. A sample holder can be mounted on the bottom of this support structure, which is located in the homogeneous region of the magnetic field.

The insert is cooled using both liquid nitrogen and two cryocoolers. A liquid nitrogen tank is located at the lowest part of the insert. This tank is screwed to an aluminium thermal shield that surrounds the support structure. In this way, the shield is cooled to 90-100 K. Secondly, the evaporated nitrogen gas flows through the current leads, cooling them to 90-100 K as well.

A 120 W heating element is placed on the liquid nitrogen tank to generate a steady flow of nitrogen gas.

Two cryocoolers provide cooling power to the support structure and the sample holder. The cold heads are mounted on the lid of the insert. The first stages of both cold heads are connected to the support structure, cooling it to 100 K. The second stages of the cold heads are connected to two copper leads which go down to the sample area. They can be connected to the sample holder using flexible copper cables. The cooling system was designed to reach a temperature of 10 K in the sample holder.

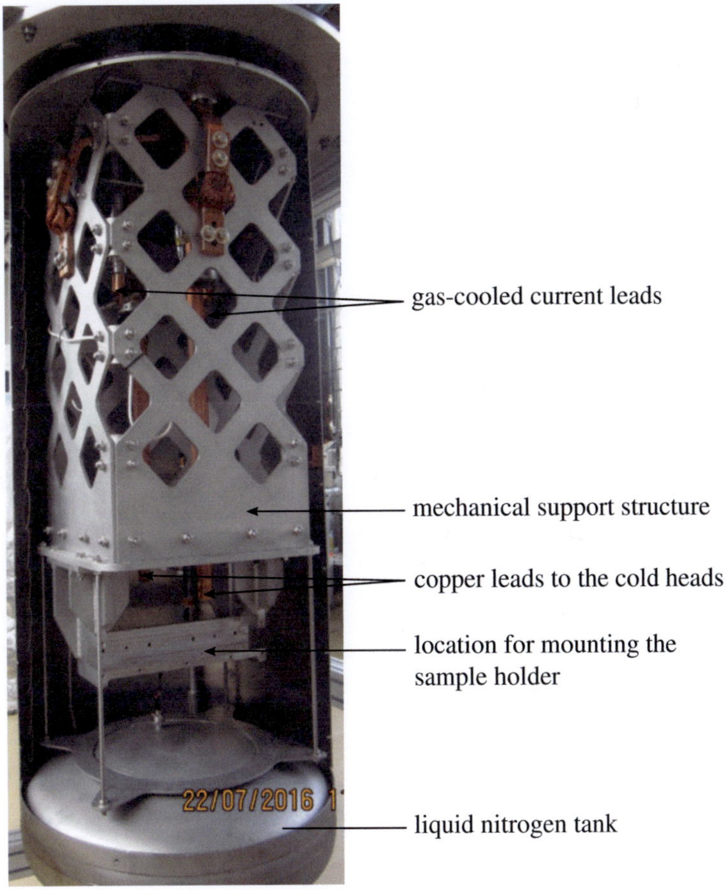

Figure 5.19: The variable temperature insert of the Vatesta facility. The aluminium thermal shield was removed before taking this picture.

A new sample holder was constructed for the Roebel cable MQE measurement. The assembled sample holder is depicted in figure 5.20 and an exploded view is shown in figure

5.21. The sample holder consists of a U-shaped and a T-shaped part, both made from glass-fibre reinforced plastic (GFRP). The U-shape has a length of 630 mm which equals five times the transposition length of the investigated cable, which is $\ell_t = 126$ mm. The T-shaped part has a length of three transpositions, and is pressed on the cable by 12 pressure springs. Forty spring contacts are built into the T-shape, enabling voltage measurements on each strand over a length of two transpositions. On both ends of the sample holder, a cable length of one transposition remains for current contacts. An 8-mm-thick copper heat spreader is attached to the bottom of the sample holder. The heat spreader as well as the current contacts are connected to the second stage of the cryocoolers using flexible copper cables.

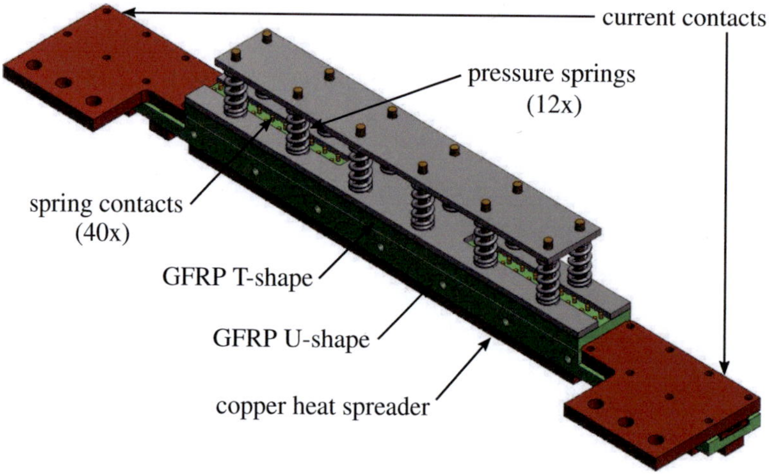

Figure 5.20: Sample holder for measuring Roebel cables on the variable temperature insert.

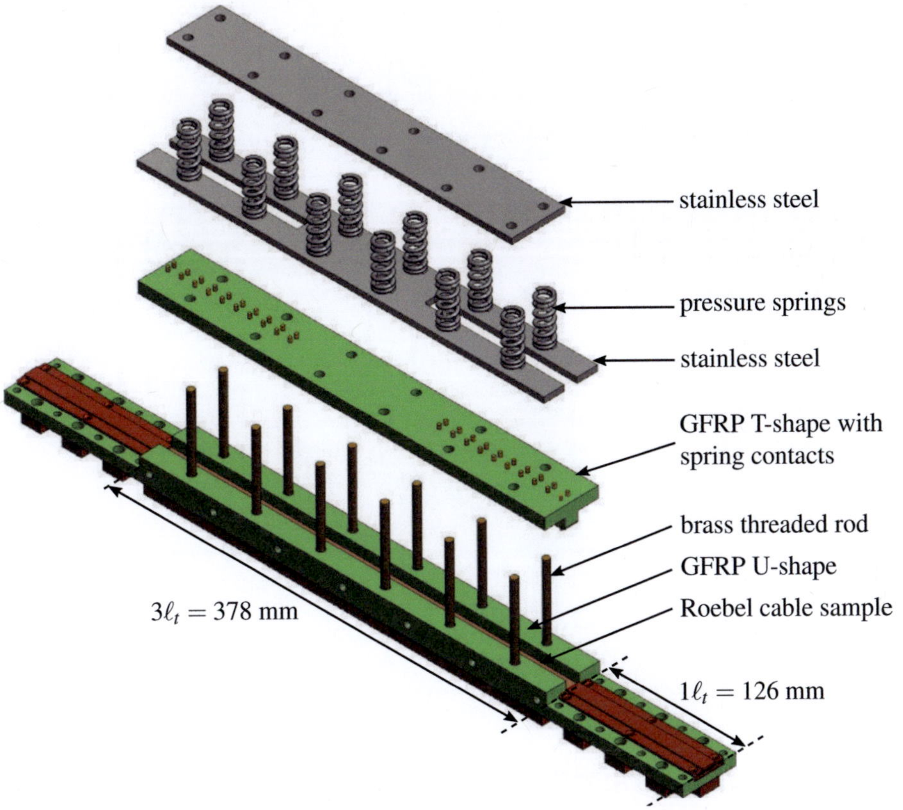

Figure 5.21: Exploded view of the sample holder with current contacts removed.

The temperature in the sample holder is controlled using a set of temperature sensors and heaters. The temperature control system is schematically shown in figure 5.22. Carbon-based TVO-sensors[2] are placed in the current contacts and in the U-shape. A Lakeshore 218 temperature controller measures the resistance of the sensors using a four-point measurement, and calculates the temperature using individual calibration curves. Four resistive heaters with a resistance of 6.8 Ω are placed on the current leads and on the heat spreader on the back of the sample holder. The heaters are controlled using a proportional-integral-derivative (PID) controller programmed in LabVIEW. The feedback loop keeps the temperature stable within 0.2 K of the set-point.

The instrumentation for the quench experiment is shown in figure 5.23. There are two current sources: one for the sample current and one for the quench heater. The sample current

[2]TVO stands for a thermally resistant, water/moisture resistant, pressurized/compacted temperature sensor. The abbreviation is translated from Russian “TBO”.

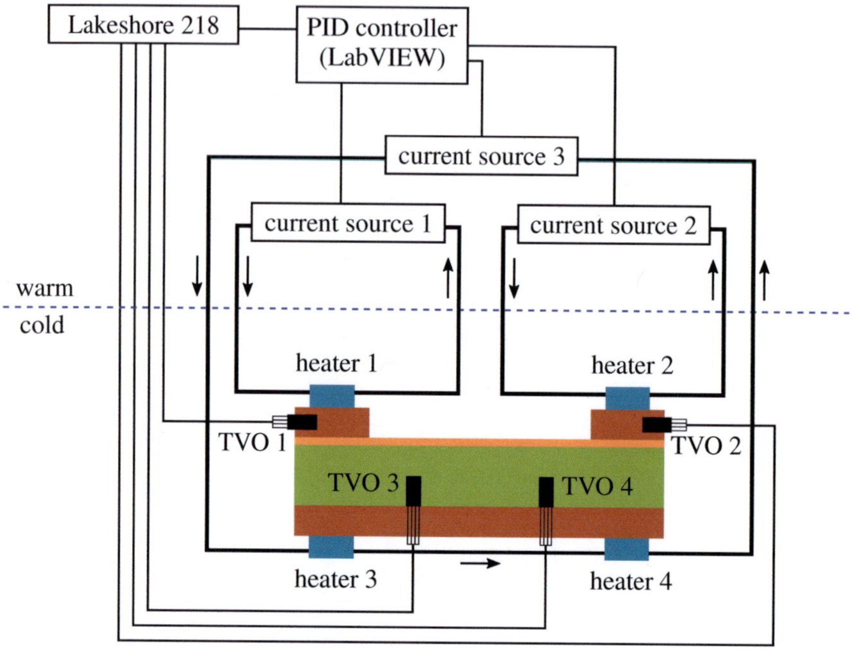

Figure 5.22: Temperature control system.

source is a Höcherl & Hackl ZS154 DC load and the heater current source is a Rohrer PA2230A amplifier. The heater consists of a layer of graphite-filled epoxy on one of the strands, through which a current is passed. The heater current is injected using a copper foil and returns through the Roebel cable. The preparation of the heater will be discussed in more detail in the next section.

The heater current source produces a considerable amount of noise. As a result, the heater would produce heat even when the set current is zero. To solve this issue, the heater is placed in series with a diode and a negative current is applied when the heater is not used. This current passes not through the heater but through another diode clamping the output of the current source. The sample current source is also placed in series with diodes. These diodes protect the sample current source from a backwards current that could be produced by the heater current source in case the Roebel cable burns out. Both current sources are placed in series with a shunt resistor of respectively 250 $\mu\Omega$ and 20 $m\Omega$ in order to measure the applied current.

The voltage is measured on each strand over a length of two transpositions. Also, the voltage over the heater and the shunt resistors is measured. All voltages signals are conditioned by isolated amplifiers and measured using a NI6289 data acquisition device. The same device controls the current sources using the analog output channels. This ensures synchronization of input and output signals.

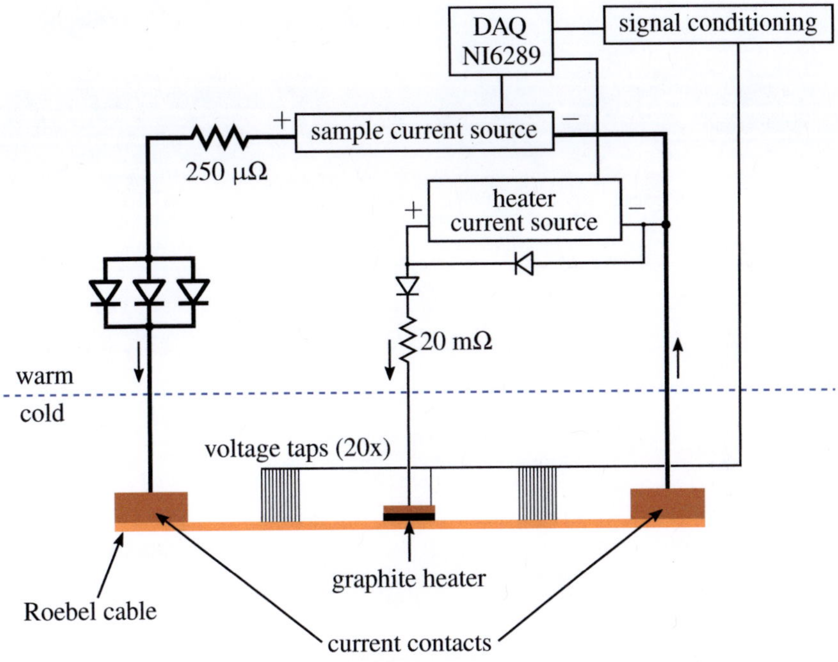

Figure 5.23: Sample instrumentation scheme.

5.6.2 Sample preparation

Ten Roebel strands of 69 cm length were prepared from 12 mm-wide SuperPower SCS12050-AP coated conductor. The strands were 5.5 mm wide and had a transposition length of 126 mm.

A graphite-based quench heater was prepared on one of the strands. The heater was placed on the cross-over, because this is the only part of the strand that can be conveniently reached from the outside. The heater was prepared as follows: First, the tape was covered with a 40 μm-thick adhesive polyimide tape. A slit of 0.3-0.5 mm width on the cross-over was left uncovered (see figure 5.24). The slit was filled with a graphite-filled epoxy (Loctite Ablestik 60L) and covered with a 10 μm-thick copper foil which functions as an electrode. The graphite-epoxy was cured at 65 °C for one hour.

After curing, the strands were assembled and the cable was mounted on the sample holder. The cable was soldered to the current contacts over one transposition length on both ends with $In_{52}Sn_{48}$. Two wires for current injection and voltage measurements were soldered to the copper electrode (figure 5.25). The neighbouring strands were also covered with polyimide tape to prevent a short circuit with the copper electrode.

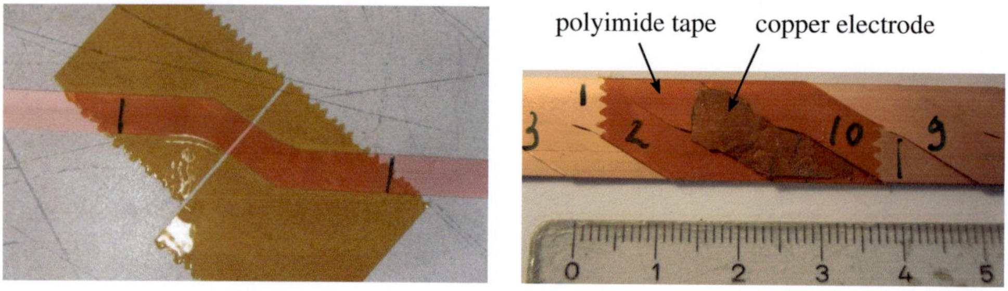

(a) (b)

Figure 5.24: (a) The slit before filling with graphite epoxy. (b) The assembled Roebel cable with the quench heater.

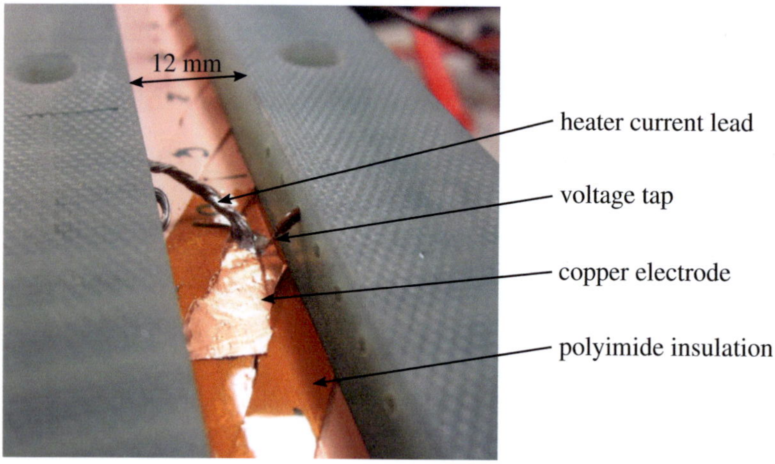

Figure 5.25: Roebel cable with quench heater in the GFRP U-shaped part of the sample holder.

An idealized cross-section of the strand with the heater is shown in figure 5.26. To induce a normal zone, a current is injected through the blue current lead and returned through the coated conductor. According to the datasheet, the graphite-epoxy has a resistivity of 50 Ωcm, which is seven to nine orders of magnitude higher than the resistivity of copper. It is thus safe to assume that heating is concentrated in the graphite-epoxy. An advantage of this type of heater is that heat is transferred quickly to the superconducting wire, because it is not insulated from the heater. Similar carbon-based heaters have been used for minimum quench energy measurements on low-temperature superconductors [50, 51, 132, 146–149] and more recently also on HTS wires [145].

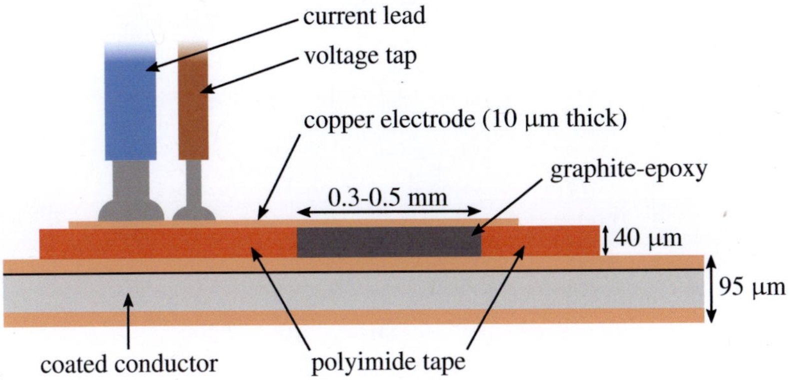

Figure 5.26: Schematic cross-section of the quench heater.

The U-shape and T-shape of the sample holder were assembled and the 12 springs were compressed by 3 mm by tightening the screws. The springs had a spring constant of 63 N/m, and therefore the total force on the sample was:

$$12 * 0.003 \text{ m} * 63 \text{ N/m} = 2268 \text{ N} \tag{5.70}$$

Considering the force is applied over three transposition lengths, the average pressure is:

$$\frac{2268 \text{ N}}{3 * 126 \text{ mm} * 12 \text{ mm}} = 0.5 \text{ MPa} \tag{5.71}$$

The force is however not homogeneously distributed, because the cable has an uneven thickness. Also, a part of the T-shape was removed by milling over a length of 4 cm near the heater, to make space for the current and voltage wires. As a result, no force is applied in this region.

The critical current of the cable was measured in a liquid nitrogen bath ($T = 77.3$ K). The voltage was measured on each strand over 252 mm using the spring contacts. The critical current was found to be 1067 A using the $E_c = 10^{-4}$ V/m criterion. For quench simulations, the critical current of the cable as a function of temperature is needed. This $I_c(T)$ relation was constructed by scaling the single tape $I_c(T)$ found in section 5.4.1 to match the critical current measured in liquid nitrogen. The resulting cable $I_c(T)$ is shown in figure 5.27. An advantage of this method is that only a single critical current measurement of the entire cable is needed. The $I_c(T)$ may however deviate at temperatures other than 77.3 K, because the magnetic self-field of a cable differs from that of a single tape.

Besides the critical current as a function of temperature, the inter-strand resistance values are needed for the calculations. The inter-strand resistance was measured in a liquid nitrogen

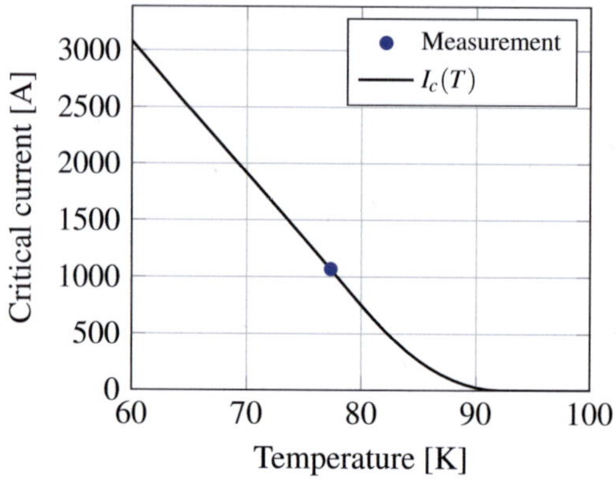

Figure 5.27: Critical current of the cable at $T = 77.3$ K and fitted temperature dependence.

bath using the method described in section 3.5. In short, a transverse current is applied between opposite strands, while measuring the potential of each strand relative to strand number 10. The inter-strand resistance parameters ρ_a and ρ_c are determined by least-squares fitting to the calculated profile. The measured profile with an applied current of 50 A as well as the least-squares fit are shown in figure 5.28. The resulting inter-strand resistance values are $\rho_a = 0.270$ µΩm and $\rho_c = 2.211$ µΩm. These values are low compared to the soldered and pressed cables investigated in chapter 3. The most likely reason for this is the fact that the cable is soldered to the copper current contacts at both ends.

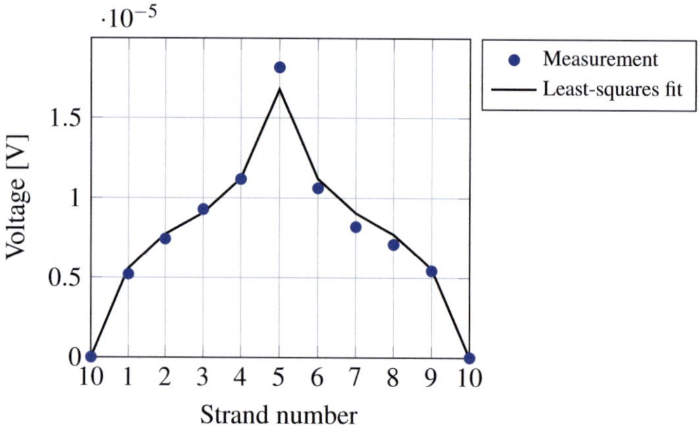

Figure 5.28: Inter-strand resistance measurement and least-squares fit. The fitting parameters are $\rho_a = 0.270$ µΩm, $\rho_c = 2.211$ µΩm. The applied transverse current was 50 A.

5.6.3 Measurement procedure

The sample holder is placed on the Vatesta insert, which is lowered into a cryostat. The air is removed from the cryostat using a vacuum pump, and then the insert is conduction cooled using a cryocooler. Because of the large mass of the insert, it takes five days before an equilibrium at about 50 K is reached. Next, the sample holder is heated to the desired measurement temperature using the four heater resistors and the PID controller. The temperature controller is left on for at least 12 hours to ensure a homogeneous temperature in the sample holder.

The current pulses applied to sample and the quench heater are shown in figure 5.29. The sample current is ramped to a constant value in 50 ms. A gradual ramp is needed to avoid inductive voltage spikes which lead to false positive quench detections. At $t = 0.2$, the quench heater is powered at a constant current for 100 ms. The sample current is stopped if the voltage over one of the strands exceeds 10 mV, or if 10 second have elapsed. In case 10 mV was reached, the measurement is registered as a quench. The 10 mV quench criterion is chosen such, because it produces no false positives, while it still stops the current quickly enough to prevent overheating. To find the MQE, the measurement is repeated with increasingly large heat pulses. A rough estimation is made using steps of 100 mJ. Subsequently, a finer increment of 10 mJ is used to find the MQE with 10 mJ precision. The waiting time between measurements is at least two minutes so that the cable can cool down sufficiently.

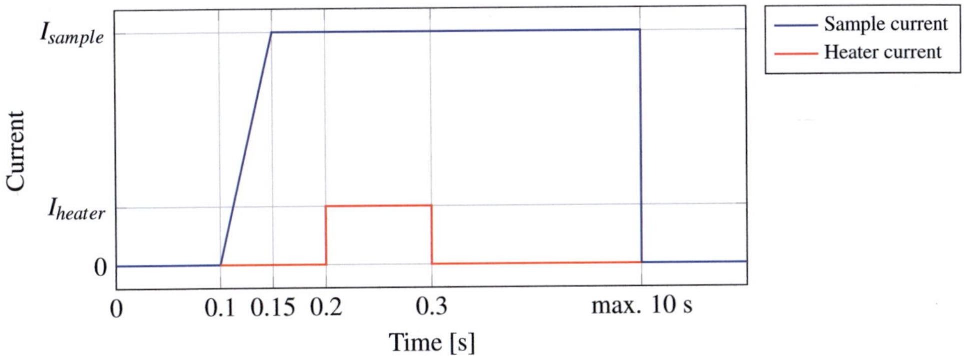

Figure 5.29: Sample and heater current pulse shapes.

5.6.4 Results

The minimum quench energy was determined in magnetic self-field at four different temperatures from 84 K decreasing to 73.5 K. Measurements at lower temperatures were not possible, as the critical current exceeds the maximum current of the power supply (1650 A). Initially, the heater energy was kept below 1 J, to avoid overheating of the heater. These measurements are shown as the closed symbols in figure 5.30. In a second set of measurements, the MQE was determined at lower applied currents, requiring an energy above 1 J (open symbols). The heater burnt out when a pulse of 2.2 J was applied at 80.5 K. Because of this, the high energy measurements at lower temperatures could not be done. The MQE of the second set is slightly lower than the first set, indicating a small degradation in critical current after more than 100 quenches.

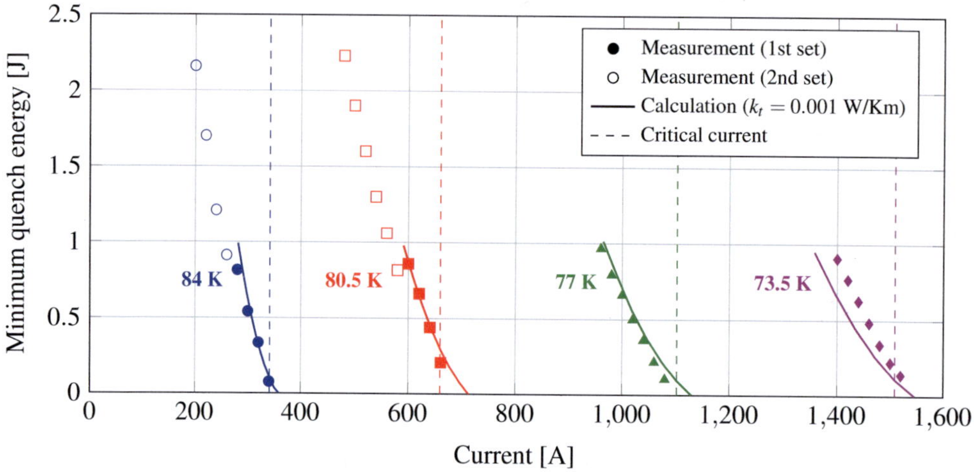

Figure 5.30: Minimum quench energy of the Roebel cable at different temperatures. The dashed lines denote the critical currents for the corresponding temperatures.

In figure 5.31, the measurements at 77.0 K are compared to the calculations. The calculations were done using the method described in section 5.3 and temperature-dependent properties as in 5.4. The calculation was done using the inter-strand resistance measured in liquid nitrogen. It also takes into account the short sample length of 630 mm. The MQE was calculated for different values of the transverse thermal conductivity k_t ranging from 0.001 to 1 W/Km. The best match with experimental data is attained for a low value of $k_t = 0.001$ W/Km. This indicates poor strand-to-strand thermal contact at the location of the heater. A likely reason is that no pressure is applied near the heater, because part of the sample holder was milled to make space for the heater wiring. Secondly, a small tension on the heater current lead could pull the tape slightly upward, creating a gap and reducing the thermal contact even further. As seen in figure 5.30, calculations with $k_t = 0.001$ W/Km make a reasonable match with experimental data at other temperatures as well.

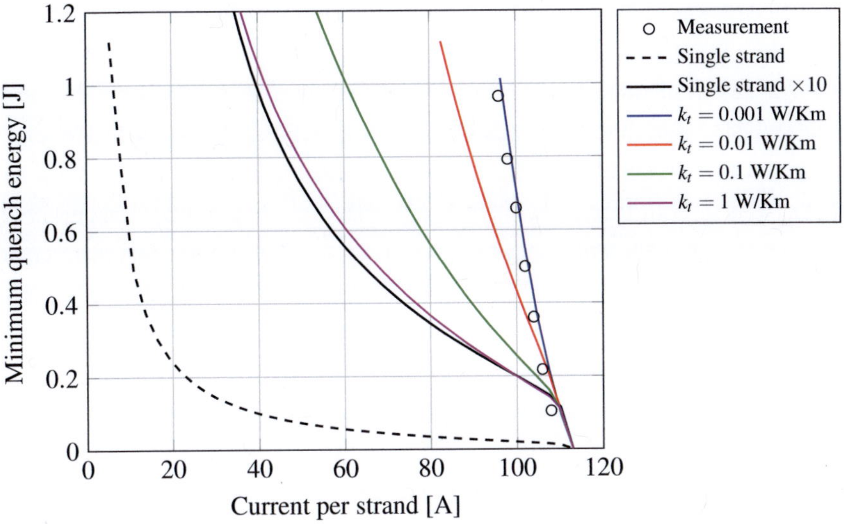

Figure 5.31: Comparison of measured and calculated minimum quench energies at $T = 77.0$ K. The cable MQE was calculated for different values of the transverse thermal conductivity k_t.

5.7 Conclusion and recommendations

The stability of Roebel cables was explored by investigating the effect of thermal and electromagnetic coupling on the minimum quench energy. The minimum quench energy was defined here as the minimum amount of energy injected into one of the strands that causes a quench of the entire cable. The calculations predict that the cable MQE equals the sum of strand MQE if the transverse thermal conductivity is high ($k_t \geq 1$ W/Km). For lower values of transverse thermal conductivity ($k_t \leq 0.1$ W/Km), the MQE depends on the inter-strand resistance, with lower values leading to a higher MQE. The most stable cables are those with which allow current redistribution, but have low transverse thermal conductivity. In this case, the hotspot is confined to a single strand, while current flows through other strands preventing further heating.

The MQE was measured in a Roebel cable at temperatures ranging from 73.5 to 84 K and in magnetic self-field. The cable was placed into a GFRP sample holder, which was conduction cooled in vacuum chamber. Because of uneven thickness, the contact area to the sample holder was small, resulting in near adiabatic conditions for short time scales. A graphite-epoxy heater was placed on one of the strands to inject heat pulses and determine the minimum quench energy. The measured MQE values were compared to calculations taking into account the measured inter-strand resistance and temperature-dependent critical current. The best match with the experimental data was found when a low transverse thermal conductivity of $k_t = 0.001$ W/Km was chosen for the calculations.

Although the investigation provides insight in the effect of current redistribution on stability, it does not suffice to make a recommendation for how to stabilize Roebel cable devices. For this, more information on the detailed conditions in such a device is needed. First, it should be known against what kind of disturbances the cable is to be stabilized. In LTS magnets, quenches can be caused by wire movements and cracks of the impregnation resin. Such disturbances, however, are unlikely to cause quenches in HTS cables, because the temperature margins and MQEs are much higher. The precise need for stabilization can only be known once actual stability problems in prototype HTS devices are encountered. Secondly, the inter-strand resistance and transverse thermal conductivities within such device should be measured for more accurate calculations.

The MQE measurement can be improved in a number of ways. First, a bare cable was used, while Roebel cable devices are likely to be impregnated for mechanical stabilization. The impregnation fills gaps between strands and in this way alters the thermal properties of the cable. For more representative MQE measurements, impregnated cable samples can be used. Secondly, the investigated sample had a rather short length of 630 mm, of which a large part of 40% is soldered in the current contacts. This led to a low average inter-strand resistance and possible improvement of stability by current redistribution over the current contacts. For more representative analysis of the stability, the use of longer samples, and preferably sub-sized prototype devices is recommended.

Chapter 6

Conclusion

In this thesis, mechanical properties and the significance of the inter-strand resistance in *RE*BCO Roebel cables were investigated. These investigations have created some fundamental knowledge that can advance the development of superconducting devices using Roebel cables or other types of HTS cables. The most important conclusions from each chapter are summarized below.

In chapter 2, the effects of transverse compression, bending and torsion on the cable were investigated. Torsion and bending take place during coil winding, while in high-field applications the Lorentz force can lead to high transverse stresses. It was found that a bare cable without impregnation is irreversible damaged by transverse stress levels as low as 40 MPa. This is much below the calculated stress levels of the dipole magnets developed in the EuCARD-2 project, which are 110 MPa for the aligned block design and 220 MPa for the cos-theta design. By impregnation, the transverse stress tolerance of two Roebel cables was increased to 253 MPa and 169 MPa. This finding was an important step for the further development of Roebel cables for use in accelerator magnets. The out-of-plane bending properties of Roebel cables were found to be similar to the bending properties of single coated conductors from which they are prepared. An explanation is that the cable geometry has only a negligible effect on the bending stress, because the strands are free to slide within the cable. One can thus predict the cable bending behaviour from the results of a single wire. The bending properties of a large number of single coated conductors were measured, in both bending directions and taking into account reversible effect. All tested conductors could be bent in compressive mode to a radius of 6 mm without permanent degradation. Finally, a torsion experiment was performed in which cable samples were twisted by screwing them in 3D-printed moulds. These moulds were made to deform the Roebel cable in the same way as in the inner winding of the EuCARD-2 cos-theta magnet. It was found that a twist pitch of 389 mm did not damage the cable. Using this result the design of the inner winding could be validated.

The second part of the thesis (chapters 3-5) concerns the inter-strand resistance and the effect on the AC loss and stability. These effects need to be understood before the inter-strand resistance can be optimized. In chapter 3, the inter-strand resistance was measured as function of transverse stress. The measurement was done by applying a transverse current between opposite strands, and measuring the voltage profile over all strands. This method is also used for Rutherford cables, which have the same transposition of strands as Roebel cables. It was found that the voltage profile can be described using a model of two parameters: ρ_a, the unit-length resistance connecting adjacent strands, and ρ_c, the unit-length resistance connecting each pair of strands. By least-square fitting to the measured voltage profile, ρ_a and ρ_c were determined in different cables. In pressed cables, inter-strand resistance values of ρ_a in the range 2.5 to 28 μΩm were found, depending on the applied pressure. The value of ρ_c was much larger than ρ_a, indicating contact between adjacent strands only. To modify the inter-strand resistance, a number of solder-filled cables was prepared. These cables had much lower inter-strand resistance values of $\rho_a = 0.19 - 0.23$ μΩm and $\rho_c = 2.4 - 2.9$ μΩm. These values, however, are not universal but depend on the cable geometry. The most important results of this chapter are the methods to measure and describe inter-strand connections.

In chapter 4, the effect of inter-strand resistance on AC loss is investigated. In particular, the induced currents through inter-strand connections are considered. These currents lead to additional dissipation known as coupling loss. A coupling loss formula was derived by simplifying the cable geometry to two parallel bars with an effective resistance. Additionally, the AC loss including contributions from both hysteresis and coupling was calculated numerically on a cross-section. Both methods take the adjacent inter-strand resistance ρ_a as an input parameter. The calculation methods were compared to AC loss measurements using the calibration-free method. A number of Roebel cables with different inter-strands resistance values were prepared. The inter-strand resistance was modified by insulating the strands, applying pressure, or filling the cable with solder. A reasonable match with the numerical model was found for magnetic field amplitudes of 5 mT and higher. The coupling loss formula overestimates the coupling loss, likely because it does not consider magnetic shielding by the superconductor. However, it can be useful to find an upper bound to the coupling loss as a function of ρ_a.

In chapter 5, the effect of the inter-strand resistance on stability is investigated. The stability is quantified using the minimum quench energy (MQE), which was defined as the smallest heat pulse applied locally to a single strand that leads to a thermal runaway of the entire cable. In superconducting cables without insulation, the parallel strands provide alternative current paths to each other. This allows current to flow around a hotspot, reducing Joule heating and potentially improving stability. A simple method to calculate the MQE taking into account current redistribution was proposed. The current and temperature distributions are described using a system of diffusion-type partial differential equations, which is solved numerically. The method takes the inter-strand resistance ρ_a and the transverse thermal conductivity k_t as parameters. A large number of MQE calculations with different parameters was done. The general trend observed is that the stability improves if the current is allowed to redistribute, but the heat is confined to the strands. This occurs if both the inter-strand resistance and the transverse thermal conductivity are low. Additionally to the calculations, a quench experiment on a short Roebel cable was done. A graphite-epoxy heater was placed on one of the strands to apply heat pulses. The cable was placed in a GFRP sample holder, which was cooled by conduction in a vacuum chamber. The MQE was measured by applying heat pulses of increasing magnitude until a quench occurred. The measurements were done with different applied currents at temperatures between 73.5 and 84 K. MQEs between 0.08 and 2.1 J were measured, decreasing to zero as the applied current approaches the critical value. The experimental results were compared with the calculated MQE values taking into account the inter-strand resistance. The best match was found if the MQE was calculated using a low thermal conductivity of 0.001 W/Km. This indicated that there was a poor thermal contact between strands.

Chapter 7

Zusammenfassung (summary in German)

Hochtemperatursupraleiterdrähte (HTS-Drähte) haben aufgrund ihrer überlegenen Performance in hohen Magnetfeldern zur Entwicklung einer neuen Generation von supraleitenden Hochfeldmagneten geführt. Mit HTS-Drähten wurde in einem vollständig supraleitenden Magneten bisher ein Rekordmagnetfeld von 27 T erreicht und Magnete mit noch höheren Feldern bis zu 32 T sind im Entstehen. HTS-Leiter werden auch für großskalige Anwendungen wie Fusionsreaktoren und Beschleunigermagnete in Betracht gezogen. Für große Spulen mit geringer Induktivität ist die Stromtragfähigkeit einzelner Supraleiterdrähte jedoch nicht ausreichend. Daher müssen Hochstromkabel verwickelt werden, die aus zahlreichen parallel verseilten Einzelleitern bestehen. Das Roebel-Kabel aus HTS Drähten der 2. Generation (*RE*BCO coated conductor) ist ein Konzept, welches am Institut für Technische Physik (ITEP) erstmals realisiert wurde und derzeit im Hinblick auf spezielle Anforderungen verschiedener Anwendungen weiterentwickelt wird. Roebel-Kabel werden hergestellt, indem meanderförmig geschnittene *RE*BCO Bandleiter zu einem Kabel verseilt werden. Roebel-Kabel besitzen einen hohen Füllgrad, eine volle Transposition der Einzelleiter und eine mechanische Biegefähigkeit, die dem einzelnen *RE*BCO-Band entspricht. Im europäischen Projekt EuCARD-2 und den Nachfolgeaktivitäten werden mehrere HTS Dipolmagnete vom Beschleunigertyp gebaut. Für zwei Designs, das "Aligned-block-coil" (CERN) und das "Cosine-theta-coil" (CEA Saclay), wurde das Roebel-Kabel wegen seiner hohen Stromdichte und der vollständigen Transposition der Bänder als Leiter ausgewählt. Das Design beider Spulen wurde ausgelegt, ein Magnetfeld von 5 T im Eigenfeld, bzw. Felder von 15.5 T ("Aligned-block-coil") und 16.9 T ("Cosine-theta-coil") in einem 13 T Hintergrundfeld zu. Diese Felder wären absolute Rekordwerte für supraleitende Dipolmagnete. Die Rolle des ITEP in diesem Projekt war die Entwicklung und Bereitstellung von Roebel-Kabeln bis 32 m Länge, die längsten bisher. Die speziellen Anforderungen von Beschleunigermagneten an das Kabel haben eine Reihe von Fragen zu den mechanischen, elektromagnetischen und thermischen Eigenschaften der Roebelkabel aufgeworfen, die die Motivation zu den Untersuchungen dieser Dissertation ergaben.

Diese Dissertation besteht aus zwei Teilen. Im ersten Teil werden die mechanischen Eigenschaften der *RE*BCO Coated Conductor und der Roebel-Kabel untersucht. Torsionen und Biegungen treten während des Spulenwickelns auf, wohingegen in Hochfeldanwendungen die Lorentzkräfte zu transversalen Spannungsbelastungen führen. Der Einfluss dieser Deformationen auf die Kabelperformance wurde mit Kurzprobenexperimenten untersucht. Es wurde festgestellt, dass das blanke Roebel-Kabel ohne jegliche Imprägnierung bereits bei transversaler Druckbelastung von nur 40 MPa irreversibel geschädigt wurde. Das ist weit unterhalb der berechneten Spannungswerte von 110 MPa für das "Aligned-block-design" und 220 MPa für das "Cosinus-theta-design" der Dipolmagnete, die im EuCARD-2 Projekt entwickelt werden. Durch Imprägnation mit Epoxidharz konnte die Widerstandsfähigkeit gegen transversale Spannung bei zwei Roebel-Kabeln auf Werte von 253 MPa und 169 MPa gesteigert werden. Dieses Ergebnis war ein sehr wichtiger Schritt für die weitere Entwicklung von Roebel-Kabeln für die Anwendung in Beschleunigermagneten. Die Biegeeigenschaften der Roebel-Kabel senkrecht zur Ebene (out-of-plane) wurden untersucht. Das Biegeverhalten der Roebelkabel ist sehr ähnlich wie das der einzelnen Bandleiter aus denen sie bestehen. Eine Erklärung dafür ist der vernachlässigbare Einfluss der Kabelgeometrie auf die Biegespannung, da die Einzelbänder im Kabel frei gleiten können. Daher kann man das Biegeverhalten des Kabels aus den Ergebnissen der Einzelbänder ableiten. Es wurden die Biegeeigenschaften einer großen Zahl einzelner und verschiedener Bandleiter in beiden Biegerichtungen vermessen, mit besonderem Augenmerk auf das reversible Verhalten. Alle getesteten Leiter konnten im kompressiven Bereich bis zu einem Biegeradius von 6 mm ohne permanente Schädigung gebogen werden. Abschließend wurde ein Torsionsexperiment durchgeführt, bei dem Kabelproben durch Eindrehen in eine 3D-gedruckte Form tordiert wurden. Die 3D-Formen dazu wurden derart ausgeführt, dass das Roebel-Kabel in die identische Torsion verformt wurde wie sie in der inneren Wicklung des EuCARD-2-cosine-theta-Magneten vorliegt. Als Ergebnis wurde kein Einfluss des erzwungenen Twistpitches von 389 mm festgestellt und damit die Machbarkeit des Magnetdesigns für die kritischste innere Windung validiert.

Der zweite Teil der Dissertation befasst sich mit Auswirkungen des elektrischen Widerstands zwischen den Einzelleitern (Strands) auf die Kabeleigenschaften. Roebel-Kabel mit geringem Widerstand zwischen den Strands gestatten Stromumverteilung entlang des

Kabels und verbessern damit die thermische Stabilisierung durch das Angebot von alternativen Strompfaden an Schwachstellen im Leiter. Niedrige Kopplungswiderstände gestatten aber auch Kopplungsströme die durch zeitveränderliche Magnetfelder angeworfen werden. Diese Kopplungsströme fließen durch die resistiven Grenzflächen zwischen den Strands und verursachen elektrische Energieverluste, die man als Kopplungsverluste bezeichnet. Kopplungsverluste sind unerwünscht da durch die entstehende Erwärmung Stabilitätsprobleme auftreten können und zusätzlich Kühlleistung nötig ist. Der Einfluss des Widerstands zwischen den Strands sowohl auf die elektrische und thermische Stabilisierung als auch auf die Wechselstromverluste muss verstanden werden, bevor der Widerstand auf den besten Kompromiss optimiert werden kann.

Es wurde ein neuer Versuchsaufbau erstellt, um den Widerstand zwischen den Strands als Funktion eines auferlegten transversalen Pressdrucks zu messen. Der Widerstand zwischen den Strands wurde durch Beaufschlagen eines Stroms zwischen den beiden gegenüberliegenden Strands und der Vermessung der Spannungsprofile zwischen allen Strandpaaren bestimmt. Die Spannungsprofile können mithilfe eines Modells beschrieben werden und der Definition zweier Parameter: ρ_a, der längennormierte Widerstand zwischen nebeneinanderliegenden, sich direkt berührenden Strands, und ρ_c, der längennormierte Widerstand zwischen beliebigen Strandpaaren. Mittels eines Least-square-Fits an den gemessen Spannungsprofilen wurden ρ_a und ρ_c in verschiedenen Kabeln bestimmt. In gepressten Kabeln wurden Widerstände ρ_a im Bereich von 2.5 bis 28 $\mu\Omega$m gefunden, abhängig vom Pressdruck. Die Werte für ρ_c waren in den gepressten Kabeln viel größer als ρ_a, ein Hinweis auf auf einen guten Kontakt lediglich zwischen benachbarten Strands. Um den Widerstand zwischen den Strands zu modifizieren wurde eine Anzahl verlöteter Kabel präpariert. Diese Kabel besaßen einen viel niedrigeren Widerstand von $\rho_a = 0.19 - 0.23$ $\mu\Omega$m und $\rho_c = 2.4 - 2.9$ $\mu\Omega$m zwischen den Strands. Die erhaltenen Messwerte sind nicht allgemein gültig sondern hängen von der Kabelgeometrie ab. Mit dieser Methode der Widerstandsmessung zwischen den Strands und der Modellierung werden die verlässlichsten Ergebnisse für die Strandkopplung gewonnen.

Der Einfluss des Widerstands zwischen den Strands auf die AC-Kopplungsverluste wurde in einem senkrecht zum Kabel orientierten sinusförmigen Magnetfeld untersucht. Eine Formel zur Beschreibung der Kopplungsverluste wurde durch Vereinfachung der Kabelgeometrie in zwei parallele Streifen mit Definition eines effektiven Widerstands erhalten. Zusätzlich wurden die Wechselstromverluste inklusive der Beiträge von sowohl Hysterese- als auch Kopplungsverlusten numerisch für den Kabelquerschnitt berechnet. Beide Methoden verwenden den Widerstandswert ρ_a benachbarter Strands als Eingangsparameter. Die berechneten Werte wurden mit Wechselstromverlustmessungen mittels einer kalibrationsfreien Methode verglichen. Es wurden eine Reihe Roebel-Kabel mit unterschiedlichen Kopplungswiderstandswerten hergestellt. Der Kopplungswiderstand wurde durch elektrische Isolation der Strands, Anwendung von

Pressdruck oder Verlöten modifiziert. Eine akzeptable übereinstimmung mit dem numerischen Modell wurde für Magnetfeldamplituden von 5 mT oder höher gefunden. Die angewandte Formel für Kopplungsverluste schätzt die Kopplungsverluste zu hoch ein, vermutlich durch Nichtberücksichtigung der Abschirmeffekte des Supraleiters. Sie macht jedoch für die Abschätzung einer Obergrenze für die Kopplungsverluste als Funktion von ρ_a Sinn.

Abschließend wurde der Einfluss des Kopplungswiderstands der Strands auf die Stabilisierung des Kabels untersucht. Die Stabilisierung wird quantifiziert durch die minimale Quenchenergie (MQE), die durch den kleinsten lokal erzeugten Wärmepuls in einem Strand definiert ist, der zu einem Quench des gesamten Kabels eskaliert. Es wurde eine einfache Methode zur Berechnung der MQE unter Berücksichtigung der Stromumverteilung vorgeschlagen. Die Strom- und Temperaturverteilungen werden dabei mit einem Satz partieller Differentialgleichungen des Diffusionstyps beschrieben, die numerisch gelöst wurden. Diese Methode benutzt den Kopplungswiderstand der Strands ρ_a und die transversale thermische Leitfähigkeit als Parameter. Es wurde eine große Zahl von MQE Berechnungen mit verschieden Parametern durchgeführt. Es ergab sich der allgemeine Trend einer verbesserten Stabilisierung wenn eine Stromumverteilung möglich ist, die Wärme verbleibt dabei jedoch am Strand. Das geschieht wenn sowohl der Kopplungswiderstand als auch die transversale Wärmeleitung klein sind. Zusätzlich zu den Berechnungen wurde ein Experiment an kurzen Roebel-Kabeln durchgeführt. Ein Graphit-Epoxidharz-Heizer wurde an einem Strand platziert um Wärmepulse zu erzeugen. Das Kabel war in einem GFK-Probenhalter im Vakuum und wurde durch Wärmeleitung gekühlt. Die MQE wurde bestimmt durch eingekoppelte Wärmepulse mit zunehmender Intensität bis ein Quench auftrat. Die Messungen wurden mit unterschiedlichen aufgeprägten Strömen durchgeführt, bei Temperaturen zwischen 73.5 und 84 K. MQE-Werte zwischen 0.08 und 2.2 J wurden gemessen, die auf Null abnahmen wenn sich der auferlegte Strom dem kritischen Wert näherte. Die experimentellen Resultate wurden unter Einbezug der Kopplungswiderstände mit den berechneten MQE-Werten verglichen. Die beste Übereinstimmung ergab sich mit einer niedrigen thermischen Wärmeleitung von 0.001 W/Km. Das ist ein Hinweis auf einen sehr geringen thermischen Kontakt zwischen den Strands im Kabel.

Appendix A

Punching quality

In figure A.1, a cross-sectional sketch of the punching tool is shown. The core of the tool consists of three parts: the stripper, the punch and the die. The punch and die are two closely fitting parts made using high-precision spark erosion. They cut the tape similar to a pair of scissors. The ideal clearance between punch and die depends on the thickness and tensile strength of the punched material. As a rule of thumb, the clearance should be 2 to 5% of the tape thickness [150, p. 42]. The job of the stripper is to keep the tape in place while it is being cut.

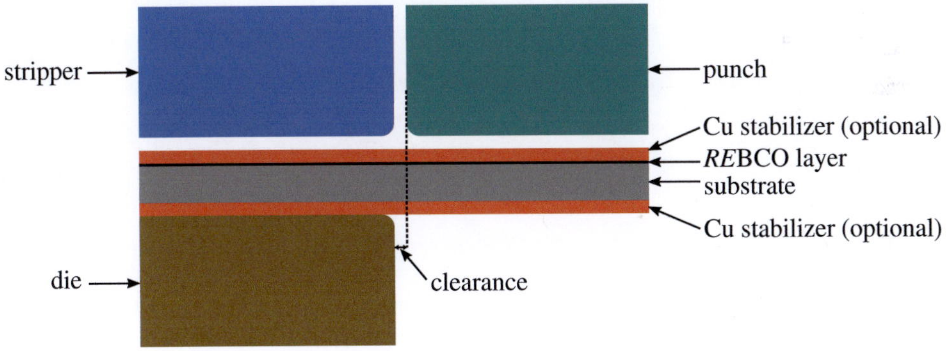

Figure A.1: Cross-sectional view of the core of a punching tool with a *RE*BCO coated conductor. The buffer stack and silver layer in the coated conductor are omitted for clarity.

The steps of a single punching operation are illustrated in figure A.2. First, a downward force is applied to the tape to keep it in place (1). Then the punch cuts the tape (2) and moves back to its original position (3). Finally, the tape is released (4).

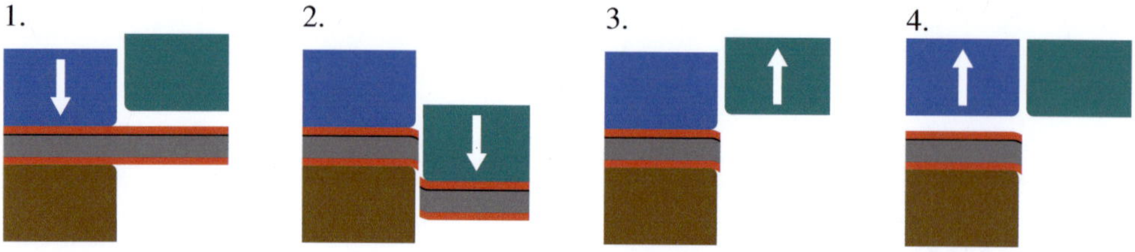

Figure A.2: The punching process in four steps. 1: A downward force is applied to the tape to keep it in place. 2: The punch moves down and cuts the tape. 3: The punch moves back. 4: The tape is released.

After each punching step, the tape is moved using a roll feeder. The transport system including feeder and reels is shown in figure A.3. The precision of the transport system is important because it determines the transposition length. Errors in transposition accumulate over the cable length and lead to problems in long-length cable as will be discussed further on.

Figure A.3: Picture of the punching tool including pneumatic press and reel-to-reel transport system.

Quality of the punched edge

Since the cutting edges of the tool are not ideally sharp, they will always leave a burr. The size of the burr depends on the clearance between punch and die, the sharpness of the cutting edges and tape properties, in particular the thickness and shear strength. We have observed a very large burr several times, with heights up to 200 μm from the tape surface (figure A.4). A large burr does not necessarily affect the critical current, but they are undesirable for various other reasons: the sharp edges may cut into the strand below, making the cable more sensitive to transverse stress. Also, the burr can physically separate the tape from the tape below, leading to high inter-strand resistance and uncontrollable variations in cable thickness. Furthermore, deformation at the edge can cause cracks in the superconducting layer, which is a brittle ceramic. These cracks can propagate throughout the rest of the tape, leading to delamination and severe degradation of the current carrying properties. In figure A.5, a delamination is shown which was observed in a Fujikura tape after punching.

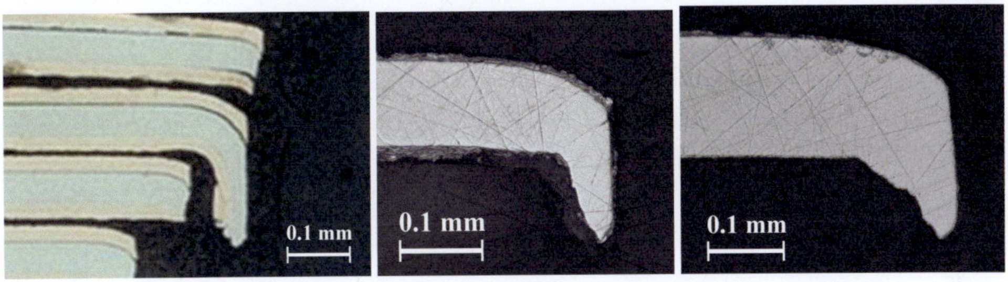

(a) (b) (c)

Figure A.4: Examples of exceptionally large punching burrs in (a) a Roebel cable from SuperPower material for CERN (image by G. Kirby), and punched stainless steel tapes of (b) 100 μm and (c) 150 μm thickness.

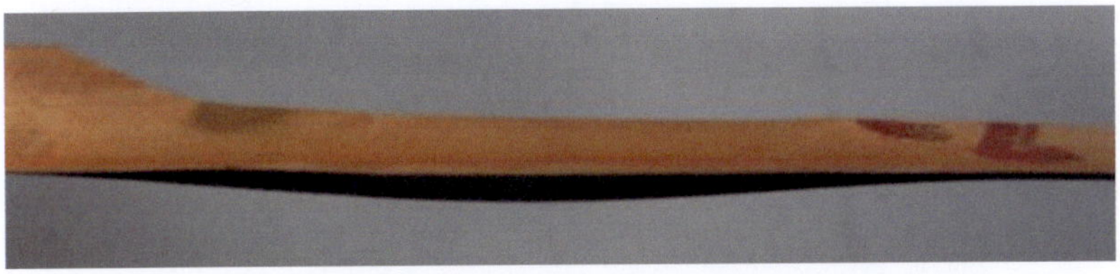

Figure A.5: Delamination in a Fujikura tape that appeared after punching.

Punching defects such as the ones described above are unacceptable for long-length production of Roebel cables. Therefore, effort was done in order to prevent such quality issues. The most important development in this aspect has been the punch-and-coat route. With this method, the strands are stabilized with copper only after punching. This route has several advantages regarding punching quality. First, punching before copper plating means that no dog-boning[1] is present during the punching process, which could hinder the function of the stripper. Secondly, unstabilized tapes have a more uniform tensile strength since the mechanical properties are dominated by the substrate material. This makes it possible to better adjust the punching tool to

[1] Dog-boning is the term commonly used to describe an uneven thickness of the copper layer. Electroplated copper layers are often thicker near the tape edges, leading to a shape that resembles a bone.

the punched material. Another advantage is that by adding copper after punching, the punched edge is fully enclosed in copper. The copper layer covers the burr, reducing its sharpness, and makes a delamination as seen in figure A.5 less likely. Furthermore, the copper enclosure protects the superconducting layer from harmful chemicals that are used in coil construction, such as soldering fluxes and impregnation resins.

The first punch-and-coat Roebel strands were prepared in a collaboration of ITEP and Bruker. A 12 mm wide coated conductor was prepared by Bruker HTS and delivered to ITEP. The tape was then punched with a 226 mm transposition length and sent back to Bruker for copper electroplating. The critical current was measured over 9 segments at three times: before punching, after punching and after copper plating. The results of these measurements are shown in figure A.6. The critical current per unit width was found to be 10.8 ± 0.2 A/mm before punching, 10.3 ± 0.3 A/mm after punching, and 10.2 ± 0.3 A/mm after adding copper. The degradation over the whole process was small amounting 6.0% on average.

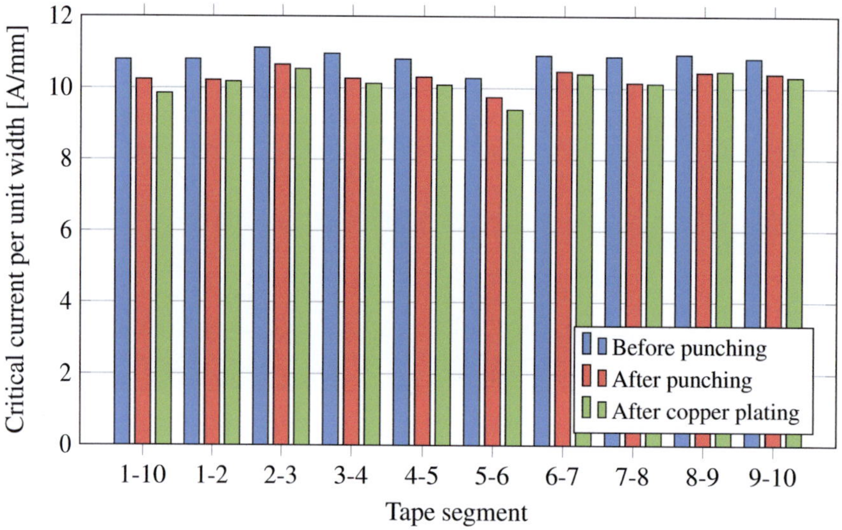

Figure A.6: Critical current per unit width before punching, after punching, and after copper plating of a Roebel strand made by the punch-and-coat process. 10 voltage taps were attached to the cable with 126 mm interval. The critical current of the nine segments and the entire cable was then determined by applying a transport current to reach a voltage criterion of $E_c = 10^{-4}$ V/m. All measurements were done in a liquid nitrogen bath ($T = 77$ K).

This route of producing copper enclosed Roebel strands has become the standard for Roebel cables in the EuCARD-2 project [151]. The method was used for the 25 m long cables for the aligned block coil and cos-theta coil, which are part of that project.

Errors in geometry

The transposition length is defined by a feeder system that moves the coated conductor through the punching tool. Although a precision of ±50 μm is stated in the datasheet, the feeder system was found to produce errors in transposition length up to 2.5 mm (see table A.1). Such errors are unimportant for short cables, but become problematic in longer lengths as the errors accumulate. In figure A.7, difficulties in winding tests of the EuCARD-2 accelerator magnets are shown. In both coils, the gap between two cross-overs became zero, due to errors in transposition length and unequal strand shift during coil winding. The unequal strand shift can also lead to increased stress concentrations under a transverse force [152].

Table A.1: Errors in transposition length measured on several punched Roebel strands.

CC material	transposition length [mm]	average error [mm]	largest error [mm]
SuperPower	226	-0.01	1.84
Bruker	226	-0.03	0.14
SuperOx	126	0.10	0.54
SuperOx	226	0.34	1.06
SuperOx	426	0.46	2.32
SuperPower	126	0.06	0.48
SuperPower	226	0.21	2.48
SuperPower	426	-0.03	2.02

In order to reduce the error in transposition length, an optical correction system for the feeder rolls has been installed. The system consists of a Keyence 2D optical micrometer and a feedback loop to the feeder rolls. The micrometer records the location of a corner of the punched coated conductor, just as it exits the punching tool. When another transposition length is punched, the next corner should be moved to exactly the same location as the previous one. If the corner is in the wrong position, this is recorded by the micrometer, and automatically corrected using the feeder rolls. This system has improved the precision in transposition length and enabled the assembly of a 100-m-long dummy cable [153].

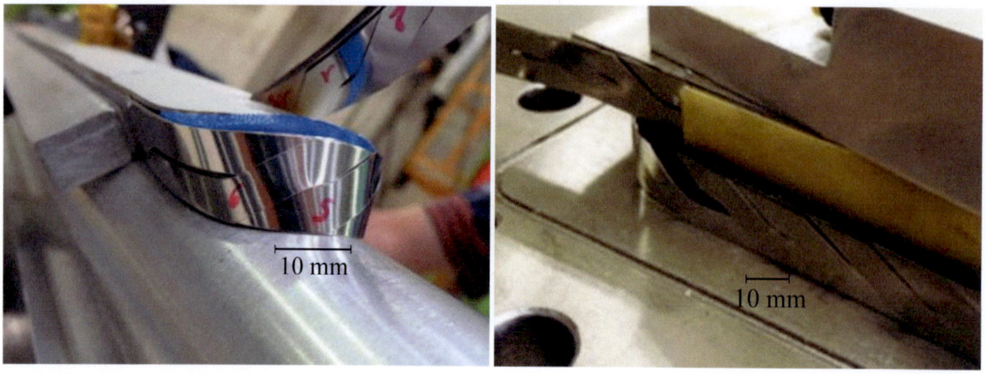

(a) **(b)**

Figure A.7: (a) Winding test using a stainless steel dummy cable of the cos-theta coil developed at CEA Saclay (picture by M. Durante). Two Roebel cross-overs are seen touching due to strand shift. (b) Similar effect observed in aligned block dummy winding at CERN (picture from G. Kirby). Errors in transposition length and strand shift have caused a “pop-up" of one of the strands.

Appendix B

Integrals

This appendix concerns the evaluation of the following integral, which is needed in section 4.3.1:

$$K_{ij} = -\frac{1}{2}\int_{y_{L,j}}^{y_{R,j}} \ln\left((y_i - y')^2 + (z_i - z_j)^2\right) \mathrm{d}y' \tag{B.1}$$

In the case of $z_i = z_j$, the second term in the logarithm equals zero and the integral becomes:

$$K_{ij} = -\int_{y_{L,j}}^{y_{R,j}} \ln|y_i - y'|\mathrm{d}y' \tag{B.2}$$

$$= -\Big[-(y_i - y')\left(\ln|y_i - y'| - 1\right)\Big]_{y'=y_{L,j}}^{y_{R,j}} \tag{B.3}$$

$$= \Big((y_i - y_{R,j})\big(\ln|y_i - y_{R,j}| - 1\big) - (y_i - y_{L,j})\big(\ln|y_i - y_{L,j}| - 1\big)\Big) \tag{B.4}$$

In the case of $z_i \neq z_j$, the integral can be evaluated by making a substitution with $u = y_i - y'$ and $\mathrm{d}u = -\mathrm{d}y'$ in equation B.1:

$$K_{ij} = \frac{1}{2}\int_{y_i - y_{L,j}}^{y_i - y_{R,j}} \ln\left(u^2 + (z_i - z_j)^2\right) \mathrm{d}u \tag{B.5}$$

and then integrating by parts:

$$v = \ln\big(u^2 + (z_i - z_j)^2\big) \tag{B.6}$$

$$\mathrm{d}v = \frac{2u}{u^2 + (z_i - z_j)^2}\mathrm{d}u \tag{B.7}$$

$$\int \ln\left(u^2 + (z_i - z_j)^2\right) \mathrm{d}u = uv - \int u\mathrm{d}v \tag{B.8}$$

$$= u \ln\left(u^2 + (z_i - z_j)^2\right) - \int \frac{2u^2}{u^2 + (z_i - z_j)^2} \mathrm{d}u \tag{B.9}$$

$$= u \ln\left(u^2 + (z_i - z_j)^2\right) - 2 \int \left(1 + \frac{u^2 - (u^2 + (z_i - z_j)^2)}{u^2 + (z_i - z_j)^2}\right) \mathrm{d}u \tag{B.10}$$

$$= u \left(\ln\left(u^2 + (z_i - z_j)^2\right) - 2\right) + 2 \int \frac{(z_i - z_j)^2}{u^2 + (z_i - z_j)^2} \mathrm{d}u \tag{B.11}$$

$$= u \left(\ln\left(u^2 + (z_i - z_j)^2\right) - 2\right) + 2(z_i - z_j) \tan^{-1}\left(\frac{u}{z_i - z_j}\right) \tag{B.12}$$

To find K_{ij}, the integration boundaries of equation B.5 are applied:

$$K_{ij} = \frac{1}{2} \left[u \left(\ln\left(u^2 + (z_i - z_j)^2\right) - 2\right) + 2(z_i - z_j) \tan^{-1}\left(\frac{u}{z_i - z_j}\right)\right]_{u = y_i - y_{L,j}}^{y_i - y_{R,j}} \tag{B.13}$$

The full expression for K_{ij} is thus:

$$K_{ij} = \begin{cases} (y_i - y_{R,j})\left(\ln|y_i - y_{R,j}| - 1\right) - (y_i - y_{L,j})\left(\ln|y_i - y_{L,j}| - 1\right) & z_i = z_j \\ \left[u \left(\frac{1}{2} \ln\left(u^2 + (z_i - z_j)^2\right) - 1\right) + (z_i - z_j) \tan^{-1}\left(\frac{u}{z_i - z_j}\right)\right]_{u = y_i - y_{L,j}}^{y_i - y_{R,j}} & z_i \neq z_j \end{cases} \tag{B.14}$$

The partial derivatives of K_{ij} with respect to y_i and z_i are needed to evaluate equation 4.68:

$$\frac{\partial K_{ij}}{\partial y_i} = \frac{1}{2} \ln\left(\frac{(y_i - y_{R,j})^2 + (z_i - z_j)^2}{(y_i - y_{L,j})^2 + (z_i - z_j)^2}\right) \tag{B.15}$$

$$\frac{\partial K_{ij}}{\partial z_i} = \begin{cases} \tan^{-1}\left(\frac{y_i - y_{R,j}}{z_i - z_j}\right) - \tan^{-1}\left(\frac{y_i - y_{L,j}}{z_i - z_j}\right) & z_i \neq z_j \\ 0 & z_i = z_j \end{cases} \tag{B.16}$$

Appendix C

Minimum propagating zones in 1D conductors

A method to construct minimum propagating zones is described by Wilson [109, p. 79]. The method is based on finding steady state solutions of the heat equation, which form the boundary between propagating and collapsing normal zones. The minimum quench energy is calculated by finding the lowest amount of energy to generate such a steady state solution. The heating is calculated following the linear current sharing model:

$$E = \begin{cases} 0 & I \leq I_c \\ \frac{\rho_e}{A}(I - I_c) & I \geq I_c \end{cases} \tag{C.1}$$

In this equation, ρ_e is the effective resistivity with contributions of all metallic layers, and A is the cross-sectional area of the conductor. The critical current is assumed to decrease linearly with temperature, while all other properties are temperature independent. These assumptions make it possible to find exact solutions of the heat equation.

Consider a wire at a temperature T_0 in adiabatic conditions. The critical current at this temperature is $I_{c,0}$ and decreases linearly to zero at the critical temperature T_c:

$$I_c(T) = \begin{cases} I_{c,0}\left(1 - \frac{T - T_0}{T_c - T_0}\right) & T \leq T_c \\ 0 & T \geq T_c \end{cases} \tag{C.2}$$

If, due to a temperature rise, the critical current I_c drops below the applied current I, part of the current is transferred to the resistive layers. This begins when the temperature reaches the current sharing temperature T_{cs}, at which the applied current equals the critical current.

$$I_c(T_{cs}) = I_{c,0}\left(1 - \frac{T_{cs} - T_0}{T_c - T_0}\right) = I \tag{C.3}$$

$$\Rightarrow T_{cs} = T_0 + (T_c - T_0)\left(1 - \frac{I}{I_{c,0}}\right) \tag{C.4}$$

If the temperature reaches the critical temperature T_c, the critical current becomes zero, and all current flows in the resistive layers. The temperature dependency of the critical current is illustrated in figure C.1.

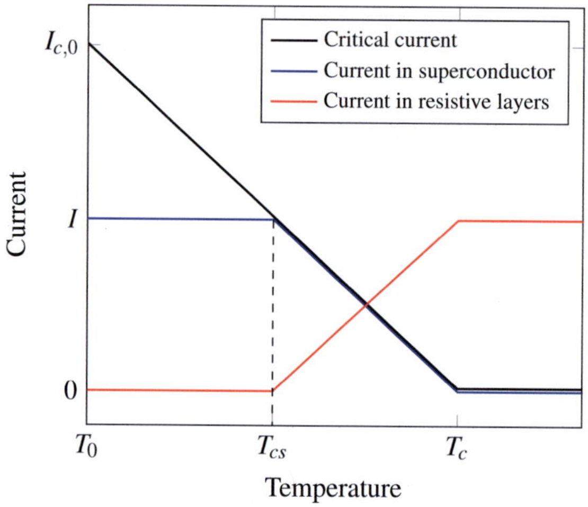

Figure C.1: Current distribution as a function of temperature following the current sharing model. The current in the superconductor is limited to the critical current. The rest of the current flows in the resistive layers. Current sharing starts at the current sharing temperature T_{cs} at which the critical current matches the applied current. At the critical temperature T_c, all current flows in the resistive layers.

The electric field equals the product of the current in the resistive layer and the effective resistance per unit length ρ_e/A:

$$E(T) = \frac{\rho_e I}{A} * \begin{cases} 0 & T \leq T_{cs} \\ \frac{T - T_{cs}}{T_c - T_{cs}} & T_{cs} \leq T \leq T_c \\ 1 & T \geq T_c \end{cases} \tag{C.5}$$

The Joule heating per unit volume is:

$$p(T) = \frac{E(T)I}{A} = \frac{\rho_e I^2}{A^2} * \begin{cases} 0 & T \leq T_{cs} \\ \frac{T-T_{cs}}{T_c-T_{cs}} & T_{cs} \leq T \leq T_c \\ 1 & T \geq T_c \end{cases} \tag{C.6}$$

Now the heat equation for a temperature profile sketched in figure C.2 is sought. In this profile, the temperature exceeds the current sharing temperature for $|x| \leq x_1$. Assuming the maximum temperature does not exceed the critical temperature, the heating is proportional to $T - T_{cs}$, as is clear from equation C.6. The point $|x| = x_2$ forms the 'cold boundary', where the temperature reaches the environmental temperature T_0. The one-dimensional heat equation for this temperature profile is:

$$\begin{cases} c\frac{\partial T}{\partial t} = k\frac{\partial^2 T}{\partial x^2} + \gamma(T - T_{cs}) & |x| \leq x_1 \\ c\frac{\partial T}{\partial t} = k\frac{\partial^2 T}{\partial x^2} & x_1 \leq |x| \leq x_2 \end{cases} \tag{C.7}$$

In this equation, c is the effective volumetric heat capacity, taking into account contributions from the different layers, k is the effective thermal conductivity and γ is the proportionality constant for the heating given by:

$$\gamma = \frac{I^2 \rho_e}{A^2(T_c - T_{cs})} \tag{C.8}$$

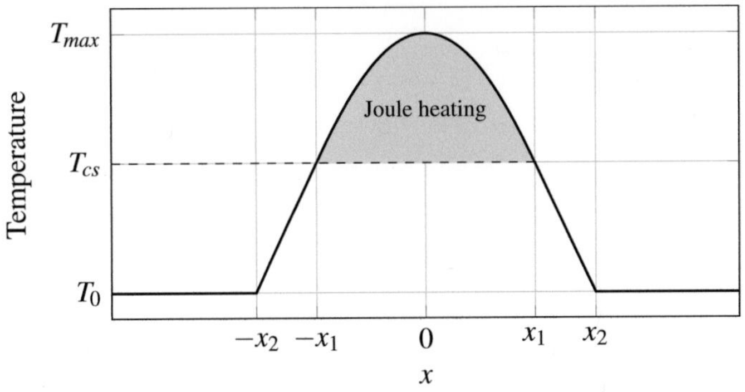

Figure C.2: Sketch of the longitudinal temperature profile of a normal zone. Joule heating occurs for $|x| \leq x_1$ where the temperature exceeds T_{cs}. The temperature remains at T_0 for $|x| \geq x_2$.

Steady state solutions fulfil the heat equation when $\partial T/\partial t = 0$. In this case, the heat equation for $|x| \leq x_1$ becomes:

$$k\frac{\partial^2 T}{\partial x^2} + \gamma(T - T_{cs}) = 0 \tag{C.9}$$

The solutions are half-period cosine functions with different amplitudes:

$$T(x) = T_{cs} + (T_{max} - T_{cs})\cos\left(\sqrt{\frac{\gamma}{k}}x\right) \tag{C.10}$$

Any value of the maximum temperature T_{max} in the range $T_{cs} < T_{max} < T_c$ yields a steady state solution. At $x = x_1$, the temperature reaches the current sharing temperature T_{cs}. The value of x_1 is therefore:

$$T(x_1) = T_{cs} = T_{cs} + (T_{max} - T_{cs})\cos\left(\sqrt{\frac{\gamma}{k}}x_1\right) \tag{C.11}$$

$$\Rightarrow x_1 = \frac{\pi}{2}\sqrt{\frac{k}{\gamma}} \tag{C.12}$$

In the range $x_1 \leq |x| \leq x_2$, the temperature does not exceed the current sharing temperature, and no Joule heating occurs. The steady state solutions should therefore satisfy:

$$\frac{\partial^2 T}{\partial x^2} = 0 \tag{C.13}$$

The steady state solutions for $x_1 \leq |x| \leq x_2$ are thus linear functions of x. The temperature profile and its derivative must be continuous at $|x| = x_1$. From equations C.12 and C.12 follows:

$$\left.\frac{\partial T}{\partial x}\right|_{x=x_1} = -(T_{max} - T_{cs})\sqrt{\frac{\gamma}{k}} \tag{C.14}$$

Therefore the temperature distribution for $x_1 \leq |x| \leq x_2$ is:

$$T(x) = T_{cs} + (T_{max} - T_{cs})\left(\frac{\pi}{2} - \sqrt{\frac{\gamma}{k}}x\right), \qquad x_1 \leq |x| \leq x_2 \tag{C.15}$$

The location of the cold boundary x_2 can now be found by equating the temperature at x_2 to the environmental temperature T_0 in above equation:

$$T(x_2) = T_0 = T_{cs} + (T_{max} - T_{cs})\left(\frac{\pi}{2} - \sqrt{\frac{\gamma}{k}}x_2\right) \tag{C.16}$$

$$\Rightarrow x_2 = \left(\frac{\pi}{2} + \frac{T_{cs} - T_0}{T_{max} - T_{cs}}\right)\sqrt{\frac{k}{\gamma}} \tag{C.17}$$

In summary, the temperature profile of the constructed normal zone is described by

$$T(x) = \begin{cases} T_{cs} + (T_{max} - T_{cs})\cos\left(\sqrt{\frac{\gamma}{k}}x\right), & |x| \leq x_1 = \frac{\pi}{2}\sqrt{\frac{k}{\gamma}} \\ T_{cs} + (T_{max} - T_{cs})\left(\frac{\pi}{2} - \sqrt{\frac{\gamma}{k}}x\right), & x_1 \leq |x| \leq x_2 = \left(\frac{\pi}{2} + \frac{T_{cs}-T_0}{T_{max}-T_{cs}}\right)\sqrt{\frac{k}{\gamma}} \\ T_0, & |x| \geq x_2 \end{cases} \tag{C.18}$$

These temperature profiles are steady state solutions everywhere except at $x = x_2$, where the temperature increases. The cold boundary thus moves away from the hotspot. As the boundary propagates outwards, the temperature gradient at $x = x_1$ decreases. This causes the Joule heating to exceed the thermal conduction away from the hot-spot, and as a result the temperature rises without bounds. The constructed temperature profiles are therefore all normal zones that lead to a quench.

To find the quench energy, the energy needed to create such propagating zone starting from the initial temperature T_0 is calculated. For $|x| \leq x_1$, the energy is:[1]

$$\varepsilon_1 = cA\int_{-x_1}^{x_1}(T(x) - T_0)\,\mathrm{d}x \tag{C.19}$$

$$= 2cA\int_0^{x_1}\left(T_{cs} - T_0 + (T_{max} - T_{cs})\cos\left(\sqrt{\frac{\gamma}{k}}x\right)\right)\mathrm{d}x \tag{C.20}$$

$$= 2cA\left(x_1(T_{cs} - T_0) + (T_{max} - T_{cs})\sqrt{\frac{k}{\gamma}}\left[\sin\left(\sqrt{\frac{\gamma}{k}}x_1\right)\right]_{x=0}^{x_1}\right) \tag{C.21}$$

$$= 2cA\left(\frac{\pi}{2}(T_{cs} - T_0) + (T_{max} - T_{cs})\right)\sqrt{\frac{k}{\gamma}} \tag{C.22}$$

The energy contained in $x_1 \leq |x| \leq x_2$ is:

$$\varepsilon_2 = 2cA * \frac{1}{2}(T_{cs} - T_0)(x_2 - x_1) \tag{C.23}$$

$$= cA(T_{cs} - T_0)\left(\left(\frac{\pi}{2} + \frac{T_{cs} - T_0}{T_{max} - T_{cs}}\right)\sqrt{\frac{k}{\gamma}} - \frac{\pi}{2}\sqrt{\frac{k}{\gamma}}\right) \tag{C.24}$$

$$= cA\sqrt{\frac{k}{\gamma}\frac{(T_{cs} - T_0)^2}{T_{max} - T_{cs}}} \tag{C.25}$$

[1] The symbol ε is used for energies to avoid confusion with the electric field E

The total energy is the sum of both contributions ε_1 and ε_2:

$$\varepsilon_{tot} = \varepsilon_1 + \varepsilon_2 = cA\sqrt{\frac{k}{\gamma}}\left(\pi(T_{cs} - T_0) + 2(T_{max} - T_{cs}) + \frac{(T_{cs} - T_0)^2}{T_{max} - T_{cs}}\right) \tag{C.26}$$

The energy depends on the value of maximum temperature T_{max}. To find the minimum quench energy, the total energy ε_{tot} is minimized with respect to T_{max}.

$$\frac{\partial \varepsilon_{tot}}{\partial T_{max}} = cA\sqrt{\frac{k}{\gamma}}\left(2 - \frac{(T_{cs} - T_0)^2}{(T_{max} - T_{cs})^2}\right) = 0 \tag{C.27}$$

$$\Rightarrow T_{max} = T_{cs} + \frac{T_{cs} - T_0}{\sqrt{2}} \tag{C.28}$$

Substituting this result into equation C.26 for the total energy gives the minimum quench energy (MQE):

$$\text{MQE} = cA\sqrt{\frac{k}{\gamma}}\left(\pi(T_{cs} - T_0) + 2\frac{T_{cs} - T_0}{\sqrt{2}} + \sqrt{2}(T_{cs} - T_0)\right) \tag{C.29}$$

$$= cA\sqrt{\frac{k}{\gamma}}\left(\pi + 2\sqrt{2}\right)(T_{cs} - T_0) \tag{C.30}$$

By substitution of the equation for γ (equation C.8), this can be rewritten to:

$$\text{MQE} = \frac{cA^2}{I}\sqrt{\frac{k(T_c - T_{cs})}{\rho_e}}\left(\pi + 2\sqrt{2}\right)(T_{cs} - T_0) \tag{C.31}$$

This formula for the MQE is only valid as long as the maximum temperature of the normal zone does not exceed the critical temperature, because this assumption was made in equation C.7.

Figure C.3 shows several minimum propagating zones for a typical coated conductor at 77 K calculated using the described method. The coated conductor parameters are listed in table C.1. The effective resistivity and volumetric heat capacity were calculated by taking a weighted average of the literature values for copper [120] and Hastelloy C276 [142] at 77 K. The thermal conductivity of a coated conductor with similar ratio of Hastelloy and copper was measured by Bagrets et al. [143].

Table C.1: Cable parameters used for example calculations. The effective resistivity and heat capacity is computed by taking a weighted average of the values for copper [120] and Hastelloy C276 [142] at 77 K. The effective thermal conductivity at 77 K were taken from a publication by Bagrets et al. [143].

	symbol	value	unit
Critical temperature	T_c	90	K
Environment temperature	T_0	77	K
Critical current at T_0	$I_{c,0}$	340	A
Tape width	W	12	mm
Substrate thickness (Hastelloy C276)	d_{sub}	50	µm
Copper stabilizer thickness	d_{cu}	42	µm
Effective resistivity	ρ	4.591	10^{-9} Ωm
Effective volumetric heat capacity	c	1.612	10^6 J/Km^3
Effective thermal conductivity	k	237.4	W/Km

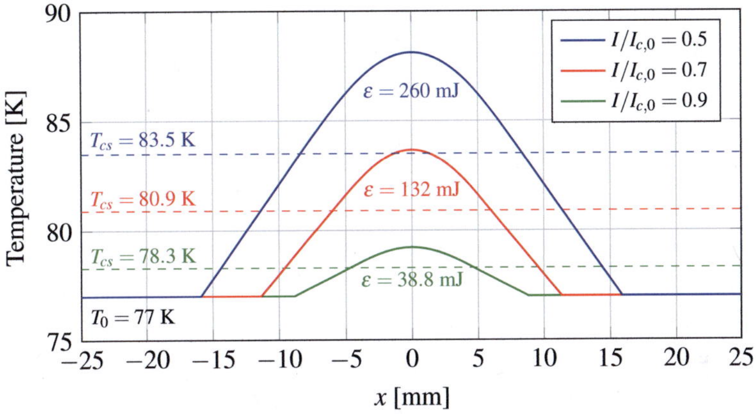

Figure C.3: Minimum propagating zones for different applied currents. The dashed lines show the corresponding current sharing temperatures according equation C.4. The energy required to create these zones from $T = T_0$ is 260, 132 and 38.8 mJ respectively for $I/I_{c,0} = 0.5$, 0.7 and 0.9.

List of symbols and abbreviations

symbol	unit	description
A	m^2	cross-sectional area
$\mathbf{A}$	Tm	magnetic vector potential (chapter 4)
A_{con}	m^2	contact area
AC		alternating current
$\mathbf{B}$	T	magnetic field vector
B_{ext}	T	external magnetic field
B_0	T	magnetic field amplitude
B_c	T	critical magnetic field
c	$JK^{-1}m^{-3}$	volumetric heat capacity
CC		coated conductor
d	m	strand thickness
d_{cu}	m	copper layer thickness
d_{sub}	m	substrate thickness
D	m	cable thickness (chapter 4)
D	-	transverse heat transfer matrix (chapter 5)
DAQ		data acquisition
$\mathbf{E}$	Vm^{-1}	electric field vector
E_c	Vm^{-1}	electric field criterion (10^{-4} V/m)
EuCARD-2		European Coordination for Accelerator Research and Development 2
f	Hz	frequency
F	N	force
GFRP		glass-fibre reinforced plastic
HTS		high-temperature superconductor

symbol	unit	description
I	A	current
I_c	A	critical current
$I_{c,0}$	A	critical current at the environmental temperature T_0
ISR		inter-strand resistance
J	Am^{-2} or Am^{-1}	current density or current per unit width
J_c	Am^{-2} or Am^{-1}	critical current density or critical current per unit width
k	$WK^{-1}m^{-1}$	thermal conductivity
k_B	JK^{-1}	Boltzmann constant ($1.381 * 10^{-23}$ JK^{-1})
k_l	$WK^{-1}m^{-1}$	longitudinal thermal conductivity
k_t	$WK^{-1}m^{-1}$	transverse thermal conductivity
L	Hm^{-1}	inductance per unit-length
LTS		low-temperature superconductor
ℓ	m	sample length
ℓ_t	m	transposition length
m	Am^{-1}	magnetic moment per unit length
M	Hm^{-1}	mutual inductance per unit length
MPZ		minimum propagating zone
MQE	J	minimum quench energy
n	-	non-linearity index
N	-	number of strands
ODE		ordinary differential equation
P	Wm^{-1}	power per unit length (chapter 4)
PDE		partial differential equation
PID		proportional-integral-derivative (controller)
q_{ext}	Wm^{-3}	external heating power per unit volume
q_{int}	Wm^{-3}	internal heating power per unit volume
q_{trans}	Wm^{-3}	transverse heat flow per unit volume
Q	Jm^{-1}	AC loss per cycle per unit length
r	Ωm^{-1}	resistance per unit length
r_e	Ωm^{-1}	effective resistance per unit length
R	m	bending radius (chapter 2)
R	Ω	resistance
R_a, R_c	Ω	adjacent and crossing elementary inter-strand resistance (Rutherford cable model)

symbol	unit	description
RACC		Roebel assembled coated conductor
*RE*BCO		rare-earth metal barium copper oxide
RRR	-	residual resistivity ratio
t	s	time
T	K	temperature
T_c	K	critical temperature
T_{cs}	K	current sharing temperature
TVO		thermally resistant, water/moisture resistant, pressurized/ compacted temperature sensor (translated from Russian “TBO”)
T_0	K	environment temperature
T_D	K	Debye temperature
U	V	electric potential
w_c	m	strand width (cross-over)
w_s	m	strand width (straight)
w_{con}	m	average width of the contact surface between strands
W	m	cable width
α	$^\circ$	cross-over angle
δ	$^\circ$	phase shift (chapter 4)
δ	K	smoothing parameter for $I_c(T)$ relation (chapter 5)
ε	-	strain (chapter 2)
ε	J	energy (chapter 5)
Λ	Hm^{-1}	inductive coupling matrix
μ_0	Hm^{-1}	vacuum permeability
ρ	Ωm	volume resistivity or unit-length resistance
ρ_a	Ωm	adjacent unit-length inter-strand resistance
ρ_c	Ωm	crossing unit-length inter-strand resistance
ρ_e	Ωm	effective resistivity
ρ_i	Ωm	ideal resistivity
ρ_{res}	Ωm	residual resistivity
σ	$(\Omega\mathrm{m})^{-1}$	inter-strand conductance matrix
τ	s	time constant
ϕ	Tm^2	magnetic flux
ω	rad s^{-1}	angular frequency

List of figures

List of tables

References

[1] H. Kamerlingh Onnes, "Further experiments with liquid helium. C. On the change of electric resistance of pure metals at very low temperatures etc. IV. The resistance of pure mercury at helium temperatures", *KNAW Proceedings* Vol. 13 No. 2 (1911)

[2] H. Kamerlingh Onnes, "Further experiments with liquid helium G. On the electrical resistance of pure metals etc. VI. On the sudden change in the rate at which the resistance of mercury disappears", *KNAW Proceedings* Vol. 14 No. 2 (1913)

[3] P. J. Lee, *Comparisons of critical and engineering current densities for superconductors available in long lengths*. URL: `https://nationalmaglab.org/magnet-development/applied-superconductivity-center/plots` (visited on 3rd July 2018)

[4] R. Perin, D. Leroy and G. Spigo, "The first, industry made, model magnet for the CERN Large Hadron Collider", *IEEE Transactions on Magnetics* Vol. 25 No. 2 (1989). DOI: `10.1109/20.92612`

[5] T. Rummel, K. Ribe, G. Ehrke, K. Rummel, A. John et al., "The Superconducting Magnet System of the Stellarator Wendelstein 7-X", *IEEE Transactions on Plasma Science* Vol. 40 No. 3 (2012). DOI: `10.1109/TPS.2012.2184774`

[6] N. Mitchell, D. Bessette, R. Gallix, C. Jong, J. Knaster et al., "The ITER Magnet System", *IEEE Transactions on Applied Superconductivity* Vol. 18 No. 2 (2008). DOI: `10.1109/TASC.2008.921232`

[7] H. Tsuji, K. Okuno, R. Thome, E. Salpietro, S. Egorov et al., "Progress of the ITER central solenoid model coil programme", *Nuclear Fusion* Vol. 41 No. 5 (2001). DOI: `10.1088/0029-5515/41/5/319`

[8] P. Ferracin, G. Ambrosio, M. Anerella, F. Borgnolutti, R. Bossert et al., “Magnet Design of the 150 mm Aperture Low-β Quadrupoles for the High Luminosity LHC”, *IEEE Transactions on Applied Superconductivity* Vol. 24 No. 3 (2014). DOI: 10.1109/TASC.2013.2284970

[9] J. G. Bednorz and K. A. Müller, “Possible high T_c superconductivity in the Ba-La-Cu-O system”, *Zeitschrift für Physik B Condensed Matter* Vol. 64 No. 2 (1986). DOI: 10.1007/BF01303701

[10] M. K. Wu, J. R. Ashburn, C. J. Torng, P. H. Hor, R. L. Meng et al., “Superconductivity at 93 K in a new mixed-phase Y-Ba-Cu-O compound system at ambient pressure”, *Physical Review Letters* Vol. 58 No. 9 (1987). DOI: 10.1103/PhysRevLett.58.908

[11] S. Yoon, J. Kim, K. Cheon, H. Lee, S. Hahn et al., “26 T 35 mm all-$GdBa_2Cu_3O_{7-x}$ multi-width no-insulation superconducting magnet”, *Superconductor Science and Technology* Vol. 29 No. 4 (2016). DOI: 10.1088/0953-2048/29/4/04LT04

[12] H. W. Weijers, W. D. Markiewicz, A. V. Gavrilin, A. J. Voran, Y. L. Viouchkov et al., “Progress in the Development and Construction of a 32-T Superconducting Magnet”, *IEEE Transactions on Applied Superconductivity* Vol. 26 No. 4 (2016). DOI: 10.1109/TASC.2016.2517022

[13] A. Sagara, H. Tamura, T. Tanaka, N. Yanagi, J. Miyazawa et al., “Helical reactor design FFHR-d1 and c1 for steady-state DEMO”, *Fusion Engineering and Design* Vol. 89 No. 9 (2014). DOI: 10.1016/j.fusengdes.2014.02.076

[14] L. Zani, C. M. Bayer, M. E. Biancolini, R. Bonifetto, P. Bruzzone et al., “Overview of Progress on the EU DEMO Reactor Magnet System Design”, *IEEE Transactions on Applied Superconductivity* Vol. 26 No. 4 (2016). DOI: 10.1109/TASC.2016.2536755

[15] M. Benedikt, D. Schulte, J. Wenninger and F. Zimmermann, “Challenges for highest energy circular colliders”, No. CERN report No. ACC-2014-0153 (2014). URL: http://cds.cern.ch/record/1742294

[16] L. Rossi, “Superconductivity: its role, its success and its setbacks in the Large Hadron Collider of CERN”, *Superconductor Science and Technology* Vol. 23 No. 3 (2010). DOI: 10.1088/0953-2048/23/3/034001

[17] L. Muzzi, L. Affinito, S. Chiarelli, V. Corato, A. della Corte et al., “Design, Manufacture, and Test of an 80 kA-Class Nb_3Sn Cable-In-Conduit Conductor With Rectangular Geometry and Distributed Pressure Relief Channels”, *IEEE Transactions on Applied Superconductivity* Vol. 27 No. 4 (2017). DOI: 10.1109/TASC.2016.2627539

[18] M. Takayasu, L. Chiesa, D. L. Harris, A. Allegritti and J. V. Minervini, "Pure bending strains of Nb_3Sn wires", *Superconductor Science and Technology* Vol. 24 No. 4 (2011). DOI: 10.1088/0953-2048/24/4/045012

[19] M. J. Wolf, W. H. Fietz, C. M. Bayer, S. I. Schlachter, R. Heller et al., "HTS CroCo: A Stacked HTS Conductor Optimized for High Currents and Long-Length Production", *IEEE Transactions on Applied Superconductivity* Vol. 26 No. 2 (2016). DOI: 10.1109/TASC.2016.2521323

[20] D. C. van der Laan, J. D. Weiss, P. Noyes, U. P. Trociewitz, A. Godeke et al., "Record current density of 344 A mm^{-2} at 4.2 K and 17 T in CORC ® accelerator magnet cables", *Superconductor Science and Technology* Vol. 29 No. 5 (2016). DOI: 10.1088/0953-2048/29/5/055009

[21] F. Gömöry, J. Šouc, M. Vojenčiak, R. Terzioglu and L. Frolek, "Design and Testing of Coils Wound Using the Conductor-On-Round-Tube (CORT) Cable", *IEEE Transactions on Applied Superconductivity* Vol. 27 No. 4 (2017). DOI: 10.1109/TASC.2016.2631340

[22] W. Goldacker, F. Grilli, E. Pardo, A. Kario, S. I. Schlachter et al., "Roebel cables from REBCO coated conductors: a one-century-old concept for the superconductivity of the future", *Superconductor Science and Technology* Vol. 27 No. 9 (2014). DOI: 10.1088/0953-2048/27/9/093001

[23] R. A. Badcock, N. J. Long, M. Mulholland, S. Hellmann, A. Wright et al., "Progress in the Manufacture of Long Length YBCO Roebel Cables", *IEEE Transactions on Applied Superconductivity* Vol. 19 No. 3 (2009). DOI: 10.1109/TASC.2009.2019065

[24] I. Rodin, V. Amoskov, E. Andreev, V. Belyakov, A. Dyomina et al., "First Experience of the HTS-II Dipole Type Magnets Development at NIIEFA", *Proceedings of the 25th Russian Particle Accelerator Conference (RuPAC2016), St. Petersburg, Russia* (2017)

[25] D. Dimos, P. Chaudhari, J. Mannhart and F. K. LeGoues, "Orientation Dependence of Grain-Boundary Critical Currents in $YBa_2Cu_3O_{7-\delta}$ Bicrystals", *Physical Review Letters* Vol. 61 No. 2 (1988). DOI: 10.1103/PhysRevLett.61.219

[26] D. T. Verebelyi, D. K. Christen, R. Feenstra, C. Cantoni, A. Goyal et al., "Low angle grain boundary transport in $YBa_2Cu_3O_{7-\delta}$ coated conductors", *Applied Physics Letters* Vol. 76 No. 13 (2000). DOI: 10.1063/1.126157

[27] M. W. Rupich, D. T. Verebelyi, W. Zhang, T. Kodenkandath and X. Li, "Metalorganic Deposition of YBCO Films for Second-Generation High-Temperature Superconductor Wires", *MRS Bulletin* Vol. 29 No. 8 (2004). DOI: 10.1557/mrs2004.163

[28] A. Usoskin and H. C. Freyhardt, "YBCO-Coated Conductors Manufactured by High-Rate Pulsed Laser Deposition", *MRS Bulletin* Vol. 29 No. 8 (2004). DOI: 10.1557/mrs2004.165

[29] X. Xiong, K. P. Lenseth, J. L. Reeves, A. Rar, Y. Qiao et al., "High Throughput Processing of Long-Length IBAD MgO and Epi-Buffer Templates at SuperPower", *IEEE Transactions on Applied Superconductivity* Vol. 17 No. 2 (2007). DOI: 10.1109/TASC.2007.899450

[30] M. W. Rupich, X. Li, C. Thieme, S. Sathyamurthy, S. Fleshler et al., "Advances in second generation high temperature superconducting wire manufacturing and R&D at American Superconductor Corporation", *Superconductor Science and Technology* Vol. 23 No. 1 (2010). DOI: 10.1088/0953-2048/23/1/014015

[31] D. Abraimov, A. Ballarino, C. Barth, L. Bottura, R. Dietrich et al., "Double disordered YBCO coated conductors of industrial scale: high currents in high magnetic field", *Superconductor Science and Technology* Vol. 28 No. 11 (2015). DOI: 10.1088/0953-2048/28/11/114007

[32] I. van Driessche, J. Feys, S. C. Hopkins, P. Lommens, X. Granados et al., "Chemical solution deposition using ink-jet printing for YBCO coated conductors", *Superconductor Science and Technology* Vol. 25 No. 6 (2012). DOI: 10.1088/0953-2048/25/6/065017

[33] H. Kutami, T. Hayashida, S. Hanyu, C. Tashita, M. Igarashi et al., "Progress in research and development on long length coated conductors in Fujikura", *Physica C: Superconductivity and its Applications* Vol. 469 No. 15 (2009). DOI: 10.1016/j.physc.2009.05.135

[34] V. R. Vlad, E. Bartolomé, M. Vilardell, A. Calleja, A. Meledin et al., "Inkjet Printing Multideposited YBCO on CGO/LMO/MgO/Y2O3/Al2 O3/Hastelloy Tape for 2G-Coated Conductors", *IEEE Transactions on Applied Superconductivity* Vol. 28 No. 4 (2018). DOI: 10.1109/TASC.2018.2808403

[35] Y. Li, L. Liu and X. Wu, "Fabrication of long REBCO coated conductors by PLD process in China", *Physica C: Superconductivity and its Applications* Vol. 518 (2015). DOI: 10.1016/j.physc.2015.03.020

[36] J. X. Jin, Y. Xin, Q. L. Wang, Y. S. He, C. B. Cai et al., "Enabling High-Temperature Superconducting Technologies Toward Practical Applications", *IEEE Transactions on Applied Superconductivity* Vol. 24 No. 5 (2014). DOI: 10.1109/TASC.2014.2346496

[37] J.-H. Lee, H. Lee, J.-W. Lee, S.-M. Choi, S.-I. Yoo et al., “RCE-DR, a novel process for coated conductor fabrication with high performance”, *Superconductor Science and Technology* Vol. 27 No. 4 (2014). DOI: 10.1088/0953-2048/27/4/044018

[38] R. Knibbe, N. M. Strickland, S. C. Wimbush, A.-H. Puichaud and N. J. Long, “Structure property relationships in a nanoparticle-free SmBCO coated conductor”, *Superconductor Science and Technology* Vol. 29 No. 6 (2016). DOI: 10.1088/0953-2048/29/6/065006

[39] S. Lee, V. Petrykin, A. Molodyk, S. Samoilenkov, A. Kaul et al., “Development and production of second generation high T_c superconducting tapes at SuperOx and first tests of model cables”, *Superconductor Science and Technology* Vol. 27 No. 4 (2014). DOI: 10.1088/0953-2048/27/4/044022

[40] X. Xiong, S. Kim, K. Zdun, S. Sambandam, A. Rar et al., “Progress in High Throughput Processing of Long-Length, High Quality, and Low Cost IBAD MgO Buffer Tapes at SuperPower”, *IEEE Transactions on Applied Superconductivity* Vol. 19 No. 3 (2009). DOI: 10.1109/TASC.2009.2018816

[41] W. Prusseit, C. Hoffmann, R. Nemetschek, G. Sigl, J. Handke et al., “Long length coated conductor fabrication by inclined substrate deposition and evaporation”, *Journal of Physics: Conference Series* Vol. 43 No. 1 (2006). DOI: 10.1088/1742-6596/43/1/054

[42] N. Glasson, M. Staines, R. Buckley, M. Pannu and S. Kalsi, “Development of a 1 MVA 3-Phase Superconducting Transformer Using YBCO Roebel Cable”, *IEEE Transactions on Applied Superconductivity* Vol. 21 No. 3 (2011). DOI: 10.1109/TASC.2010.2087738

[43] N. D. Glasson, M. P. Staines, Z. Jiang and N. S. Allpress, “Verification Testing for a 1 MVA 3-Phase Demonstration Transformer Using 2G-HTS Roebel Cable”, *IEEE Transactions on Applied Superconductivity* Vol. 23 No. 3 (2013). DOI: 10.1109/TASC.2012.2234919

[44] N. Glasson, M. Staines, N. Allpress, M. Pannu, J. Tanchon et al., “Test Results and Conclusions From a 1 MVA Superconducting Transformer Featuring 2G HTS Roebel Cable”, *IEEE Transactions on Applied Superconductivity* Vol. 27 No. 4 (2017). DOI: 10.1109/TASC.2016.2639032

[45] S. S. Fetisov, V. V. Zubko, S. Y. Zanegin, A. A. Nosov, V. S. Vysotsky et al., “Development and Characterization of a 2G HTS Roebel Cable for Aircraft Power Systems”,

IEEE Transactions on Applied Superconductivity Vol. 26 No. 3 (2016). DOI: 10.1109/TASC.2016.2549036

[46] A. Kario, M. Vojenciak, F. Grilli, A. Kling, B. Ringsdorf et al., "Investigation of a Rutherford cable using coated conductor Roebel cables as strands", *Superconductor Science and Technology* Vol. 26 No. 8 (2013). DOI: 10.1088/0953-2048/26/8/085019

[47] L. Rossi, A. Badel, M. Bajko, A. Ballarino, L. Bottura et al., "The EuCARD-2 Future Magnets European Collaboration for Accelerator-Quality HTS Magnets", *IEEE Transactions on Applied Superconductivity* Vol. 25 No. 3 (2015). DOI: 10.1109/TASC.2014.2364215

[48] G. Kirby, J. Nugteren, A. Ballarino, L. Bottura, N. Chouika et al., "Accelerator-Quality HTS Dipole Magnet Demonstrator Designs for the EuCARD-2 5-T 40-mm Clear Aperture Magnet", *IEEE Transactions on Applied Superconductivity* Vol. 25 No. 3 (2015). DOI: 10.1109/TASC.2014.2361933

[49] C. Lorin, M. Segreti, A. Ballarino, L. Bottura, M. Durante et al., "Cos-theta Design of Dipole Inserts Made of REBCO-Roebel or BSCCO-Rutherford Cables", *IEEE Transactions on Applied Superconductivity* Vol. 25 No. 3 (2015). DOI: 10.1109/TASC.2014.2360422

[50] G. P. Willering, A. P. Verweij, J. Kaugerts and H. H. J. ten Kate, "Stability of Nb-Ti Rutherford Cables Exhibiting Different Contact Resistances", *IEEE Transactions on Applied Superconductivity* Vol. 18 No. 2 (2008). DOI: 10.1109/TASC.2008.920560

[51] W. M. de Rapper, L. R. Oberli, B. Bordini, E. Takala and H. H. J. ten Kate, "Critical Current and Stability of High-J_c Nb_3Sn Rutherford Cables for Accelerator Magnets", *IEEE Transactions on Applied Superconductivity* Vol. 21 No. 3 (2011). DOI: 10.1109/TASC.2010.2103549

[52] C. Lorin, M. Durante, P. Fazilleau, G. Kirby and L. Rossi, "Development of a Roebel-Cable-Based cos θ Dipole: Design and Windability of Magnet Ends", *IEEE Transactions on Applied Superconductivity* Vol. 26 No. 3 (2016). DOI: 10.1109/TASC.2016.2528542

[53] J. van Nugteren, G. Kirby, G. de Rijk, L. Rossi, H. ten Kate et al., "Study of a 5 T Research Dipole Insert-Magnet Using an Anisotropic ReBCO Roebel Cable", *IEEE Transactions on Applied Superconductivity* Vol. 25 No. 3 (2015). DOI: 10.1109/TASC.2014.2361797

[54] T. Takao, S. Koizuka, K. Oi, A Ishiyama, J. Kato et al., "Characteristics of Compressive Strain and Superconducting Property in YBCO Coated Conductor", *IEEE Transactions on Applied Superconductivity* Vol. 17 No. 2 (2007). DOI: 10.1109/TASC.2007.899654

[55] S. Otten, M. Dhallé, P. Gao, W. Wessel, A. Kario et al., "Enhancement of the transverse stress tolerance of REBCO Roebel cables by epoxy impregnation", *Superconductor Science and Technology* Vol. 28 No. 6 (2015). DOI: 10.1088/0953-2048/28/6/065014

[56] N. Cheggour, J. Ekin, C. Thieme and Y. Xie, "Effect of Fatigue Under Transverse Compressive Stress on Slit Y-Ba-Cu-O Coated Conductors", *IEEE Transactions on Applied Superconductivity* Vol. 17 No. 2 (2007). DOI: 10.1109/TASC.2007.897918

[57] D. Uglietti, R. Wesche and P. Bruzzone, "Effect of transverse load on the critical current of a coated conductor Roebel cable", *Superconductor Science and Technology* Vol. 26 No. 7 (2013). DOI: 10.1088/0953-2048/26/7/074002

[58] L. Chiesa, N. Allen and M. Takayasu, "Electromechanical Investigation of 2G HTS Twisted Stacked-Tape Cable Conductors", *IEEE Transactions on Applied Superconductivity* Vol. 24 No. 3 (2014). DOI: 10.1109/TASC.2013.2284854

[59] J. Fleiter, A. Ballarino, L. Bottura and P. Tixador, "Electrical characterization of REBCO Roebel cables", *Superconductor Science and Technology* Vol. 26 No. 6 (2013). DOI: 10.1088/0953-2048/26/6/065014

[60] T. Takematsu, R. Hu, T. Takao, Y. Yanagisawa, H. Nakagome et al., "Degradation of the performance of a YBCO-coated conductor double pancake coil due to epoxy impregnation", *Physica C: Superconductivity and its Applications* Vol. 470 No. 17-18 (2010). DOI: 10.1016/j.physc.2010.06.009

[61] C. Barth, N. Bagrets, K. P. Weiss, C. M. Bayer and T. Bast, "Degradation free epoxy impregnation of *RE*BCO coils and cables", *Superconductor Science and Technology* Vol. 26 No. 5 (2013). DOI: 10.1088/0953-2048/26/5/055007

[62] S. Matsumoto, T. Kiyoshi, A. Otsuka, M. Hamada, H. Maeda et al., "Generation of 24 T at 4.2 K using a layer-wound GdBCO insert coil with Nb_3Sn and Nb-Ti external magnetic field coils", *Superconductor Science and Technology* Vol. 25 No. 2 (2012). DOI: 10.1088/0953-2048/25/2/025017

[63] K. Mizuno, M. Ogata and K. Nagashima, "An Innovative Superconducting Coil Fabrication Method with YBCO Coated Conductors", *Quarterly Report of RTRI* Vol. 54 No. 1 (2013). DOI: 10.2219/rtriqr.54.46

[64] U. P. Trociewitz, M. Dalban-Canassy, M. Hannion, D. K. Hilton, J. Jaroszynski et al., "35.4 T field generated using a layer-wound superconducting coil made of $(RE)Ba_2Cu_3O_{7-x}$ (RE = rare earth) coated conductor", *Applied Physics Letters* Vol. 99 No. 20 (2011). DOI: 10.1063/1.3662963

[65] Y. Yanagisawa, K. Sato, R. Piao, H. Nakagome, T. Takematsu et al., "Removal of degradation of the performance of an epoxy impregnated YBCO-coated conductor double pancake coil by using a polyimide-electrodeposited YBCO-coated conductor", *Physica C: Superconductivity and its Applications* Vol. 476 (2012). DOI: 10.1016/j.physc.2012.01.025

[66] H. Boschman, A. Verweij, S. Wessel, H. ten Kate and L. J. M. van de Klundert, "The effect of transverse loads up to 300 MPa on the critical currents of Nb_3Sn cables [for LHC]", *IEEE Transactions on Magnetics* Vol. 27 No. 2 (1991). DOI: 10.1109/20.133551

[67] S. Otten, *Transverse pressure dependence of the critical current in epoxy impregnated REBCO Roebel cables*. MSc. assignment, University of Twente (2014). URL: http://essay.utwente.nl/66576/

[68] A. Verweij, "Current redistribution in the cables of LHC magnets", *CERN LHC project note 90* (1997)

[69] P. Gao, W. Wessel, M. Dhallé, S. Otten, A. Kario et al., "Effect of tape layout and impregnation method on transverse pressure dependence of critical current in REBCO Roebel cables", *Presented at the 2016 Applied Superconductivity Conference, Denver, USA* (2016)

[70] S. Otten, A. Kario, A. Kling and W. Goldacker, "Bending properties of different REBCO coated conductor tapes and Roebel cables at T = 77 K", *Superconductor Science and Technology* Vol. 29 No. 12 (2016). DOI: 10.1088/0953-2048/29/12/125003

[71] N. Cheggour, J. W. Ekin, C. L. H. Thieme, Y.-Y. Xie, V. Selvamanickam et al., "Reversible axial-strain effect in Y–Ba–Cu–O coated conductors", *Superconductor Science and Technology* Vol. 18 No. 12 (2005). DOI: 10.1109/TASC.2005.849364

[72] A. Usoskin, A. Rutt, J. Knoke, H. Krauth and T. Arndt, "Long-length YBCO coated stainless steel tapes with high critical currents", *IEEE Transactions on Applied Superconductivity* Vol. 15 No. 2 (2005). DOI: 10.1109/TASC.2005.847669

[73] W. Goldacker, A. Frank, R. Heller, S. Schlachter, B. Ringsdorf et al., "ROEBEL Assembled Coated Conductors (RACC): Preparation, Properties and Progress", *IEEE Transactions on Applied Superconductivity* Vol. 17 No. 2 (2007). DOI: 10.1109/TASC.2007.899417

[74] M. Sugano, K. Osamura, W. Prusseit, R. Semerad, T. Kuroda et al., "Reversible strain dependence of critical current in 100 A class coated conductors", *IEEE Transactions on Applied Superconductivity* Vol. 15 No. 2 (2005). DOI: 10.1109/TASC.2005.849365

[75] H. Shin, M. Dedicatoria, J. Dizon, H. Ha and S. Oh, "Bending strain characteristics of critical current in REBCO CC tapes in different modes", *Physica C: Superconductivity and its Applications* Vol. 469 No. 15-20 (2009). DOI: 10.1016/j.physc.2009.05.067

[76] H. Shin and M. Dedicatoria, "Critical current degradation behaviors of ReBCO coated conductor tapes under pressurized liquid nitrogen", *Physica C: Superconductivity and its Applications* Vol. 470 No. 20 (2010). DOI: 10.1016/j.physc.2010.05.105

[77] H.-S. Shin, K.-H. Kim, J. R. C. Dizon, T.-Y. Kim, R.-K. Ko et al., "The strain effect on critical current in YBCO coated conductors with different stabilizing layers", *Superconductor Science and Technology* Vol. 18 No. 12 (2005). DOI: 10.1088/0953-2048/18/12/023

[78] H. S. Shin, Z. Bautista, A. Gorospe, J. H. Lee, H. Lee et al., "Electro-Mechanical Properties of Single-Side Brass Foil Laminated Coated Conductor Tapes at 77 K Under Self-Field", *IEEE Transactions on Applied Superconductivity* Vol. 26 No. 4 (2016). DOI: 10.1109/TASC.2016.2528549

[79] H.-S. Shin, A. Gorospe, Z. Bautista and M. J. Dedicatoria, "Evaluation of the electromechanical properties in GdBCO coated conductor tapes under low cyclic loading and bending", *Superconductor Science and Technology* Vol. 29 No. 1 (2016). DOI: 10.1088/0953-2048/29/1/014001

[80] M. Sugano, K. Shikimachi, N. Hirano and S. Nagaya, "The reversible strain effect on critical current over a wide range of temperatures and magnetic fields for YBCO coated conductors", *Superconductor Science and Technology* Vol. 23 No. 8 (2010). DOI: 10.1088/0953-2048/23/8/085013

[81] D. C. van der Laan and J. W. Ekin, "Large intrinsic effect of axial strain on the critical current of high-temperature superconductors for electric power applications", *Applied Physics Letters* Vol. 90 No. 5 (2007). DOI: 10.1063/1.2435612

[82] D. C. van der Laan and J. W. Ekin, "Dependence of the critical current of $YBa_2Cu_3O_{7-\delta}$ coated conductors on in-plane bending", *Superconductor Science and Technology* Vol. 21 No. 11 (2008). DOI: 10.1088/0953-2048/21/11/115002

[83] D. C. van der Laan, J. W. Ekin, J. F. Douglas, C. C. Clickner, T. C. Stauffer et al., "Effect of strain, magnetic field and field angle on the critical current density of $YBa_2Cu_3O_{7-\delta}$ coated conductors", *Superconductor Science and Technology* Vol. 23 No. 7 (2010). DOI: 10.1088/0953-2048/23/7/072001

[84] D. C. van der Laan, X. F. Lu and L. F. Goodrich, "Compact $GdBa_2Cu_3O_{7-\delta}$ coated conductor cables for electric power transmission and magnet applications", *Superconductor Science and Technology* Vol. 24 No. 4 (2011). DOI: 10.1088/0953-2048/24/4/042001

[85] D. C. van der Laan, D. Abraimov, A. A. Polyanskii, D. C. Larbalestier, J. F. Douglas et al., "Anisotropic in-plane reversible strain effect in $Y_{0.5}Gd_{0.5}Ba_2Cu_3O_{7-\delta}$ coated conductors", *Superconductor Science and Technology* Vol. 24 No. 11 (2011). DOI: 10.1088/0953-2048/24/11/115010

[86] K. Ilin, K. A. Yagotintsev, C. Zhou, P. Gao, J. Kosse et al., "Experiments and FE modeling of stress-strain state in ReBCO tape under tensile, torsional and transverse load", *Superconductor Science and Technology* Vol. 28 No. 5 (2015). DOI: 10.1088/0953-2048/28/5/055006

[87] C. Zhou, K. A. Yagotintsev, P. Gao, T. J. Haugan, D. C. van der Laan et al., "Critical Current of Various REBCO Tapes Under Uniaxial Strain", *IEEE Transactions on Applied Superconductivity* Vol. 26 No. 4 (2016). DOI: 10.1109/TASC.2016.2535202

[88] D. Uglietti, B. Seeber, V. Abächerli, W. L. Carter and R. Flükiger, "Critical currents versus applied strain for industrial Y-123 coated conductors at various temperatures and magnetic fields up to 19 T", *Superconductor Science and Technology* Vol. 19 No. 8 (2006). DOI: 10.1088/0953-2048/19/8/032

[89] M. Sugano, S. Choi, A. Miyazoe, K. Miyamatsu, T. Ando et al., "Strain Analysis of $I_c(\varepsilon)$ Characteristic of YBCO Coated Conductor Measured by a Walters Spring", *IEEE Transactions on Applied Superconductivity* Vol. 18 No. 2 (2008). DOI: 10.1109/TASC.2008.922298

[90] C. Barth, G. Mondonico and C. Senatore, "Electro-mechanical properties of REBCO coated conductors from various industrial manufacturers at 77 K, self-field and 4.2 K, 19 T", *Superconductor Science and Technology* Vol. 28 No. 4 (2015). DOI: 10.1088/0953-2048/28/4/045011

[91] W. Goldacker, J. Kessler, B. Ullmann, E. Mossang and M. Rikel, "Axial tensile, transverse compressive and bending strain experiments on Bi(2223)/AgMg single core tapes", *IEEE Transactions on Applied Superconductivity* Vol. 5 No. 2 (1995). DOI: 10.1109/77.402937

[92] W. Goldacker, S. I. Schlachter, R. Nast, H. Reiner, S. Zimmer et al., "Bending strain investigations on BSCCO(2223) tapes at 77 K applying a new bending technique", *AIP Conference Proceedings* Vol. 614 No. 1 (2002). DOI: 10.1063/1.1472575

[93] N. Cheggour, J. W. Ekin, Y.-Y. Xie, V. Selvamanickam, C. L. H. Thieme et al., "Enhancement of the irreversible axial-strain limit of Y-Ba-Cu-O-coated conductors with the addition of a Cu layer", *Applied Physics Letters* Vol. 87 No. 21 (2005). DOI: 10.1063/1.2136231

[94] M. Sugano, S. Machiya, H. Oguro, M. Sato, T. Koganezawa et al., "The effect of the 2D internal strain state on the critical current in GdBCO coated conductors", *Superconductor Science and Technology* Vol. 25 No. 5 (2012). DOI: 10.1088/0953-2048/25/5/054014

[95] J. Ekin, "Strain scaling law for flux pinning in practical superconductors. Part 1: Basic relationship and application to Nb_3Sn conductors", *Cryogenics* Vol. 20 No. 11 (1980). DOI: 10.1016/0011-2275(80)90191-5

[96] M. Sugano, T. Nakamura, T. Manabe, K. Shikimachi, N. Hirano et al., "The intrinsic strain effect on critical current under a magnetic field parallel to the c axis for a MOCVD-YBCO-coated conductor", *Superconductor Science and Technology* Vol. 21 No. 11 (2008). DOI: 10.1088/0953-2048/21/11/115019

[97] N. Bagrets, E. Weiss, S. Westenfelder and K. P. Weiss, "Cryogenic Test Facility CryoMaK", *IEEE Transactions on Applied Superconductivity* Vol. 22 No. 3 (2012). DOI: 10.1109/TASC.2011.2176902

[98] J. Lu, K. Han, W. R. Sheppard, Y. L. Viouchkov, K. W. Pickard et al., "Lap Joint Resistance of YBCO Coated Conductors", *IEEE Transactions on Applied Superconductivity* Vol. 21 No. 3 (2011). DOI: 10.1109/TASC.2010.2091934

[99] Y. Seino, S. Ito and H. Hashizume, "Joint Resistance Characteristics of Mechanical Lap Joint of a GdBCO Tape With a Change in Temperature and Magnetic Field", *IEEE Transactions on Applied Superconductivity* Vol. 24 No. 3 (2014). DOI: 10.1109/TASC.2013.2289356

[100] S. Ito, Y. Seino, T. Nishio, H. Oguro and H. Hashizume, “Structure and Magnetic Field Dependences of Joint Resistance in a Mechanical Joint of REBCO Tapes”, *IEEE Transactions on Applied Superconductivity* Vol. 26 No. 4 (2016). DOI: 10.1109/TASC.2016.2539210

[101] M. Takayasu, L. Chiesa and J. V. Minervini, “Development of Termination Methods for 2G HTS Tape Cable Conductors”, *IEEE Transactions on Applied Superconductivity* Vol. 24 No. 3 (2014). DOI: 10.1109/TASC.2013.2280839

[102] N. Bagrets, A. Augieri, G. Celentano, G. Tomassetti, K. P. Weiss et al., “Investigation of ReBCO Conductor Tape Joints for Superconducting Applications”, *IEEE Transactions on Applied Superconductivity* Vol. 25 No. 3 (2015). DOI: 10.1109/TASC.2014.2373055

[103] W. Liu, X. Zhang, Y. Liu, J. Zhou and Y. Zhou, “Lap Joint Characteristics of the YBCO Coated Conductors Under Axial Tension”, *IEEE Transactions on Applied Superconductivity* Vol. 24 No. 6 (2014). DOI: 10.1109/TASC.2014.2332444

[104] A. Grether, C. Scheuerlein, A. Ballarino and L. Bottura, “Electromechanical behaviour of REBCO tape lap splices under transverse compressive loading”, *Superconductor Science and Technology* Vol. 29 No. 7 (2016). DOI: 10.1088/0953-2048/29/7/074004

[105] J. Fleiter and A. Ballarino, “In-Field Electrical Resistance at 4.2 K of REBCO Splices”, *IEEE Transactions on Applied Superconductivity* Vol. 27 No. 4 (2017). DOI: 10.1109/TASC.2017.2659618

[106] K. Kawai, S. Ito, Y. Seino, N. Yanagi, H. Tamura et al., “Optimization of a Mechanical Bridge Joint Structure in a Stacked HTS Conductor”, *IEEE Transactions on Applied Superconductivity* Vol. 23 No. 3 (2013). DOI: 10.1109/TASC.2013.2239335

[107] V. Sytnikov, G. Svalov, S. Akopov and I. Peshkov, “Coupling losses in superconducting transposed conductors located in changing magnetic fields”, *Cryogenics* Vol. 29 No. 9 (1989). DOI: 10.1016/0011-2275(89)90207-5

[108] A. P. Verweij and H. H. J. ten Kate, “Coupling currents in Rutherford cables under time varying conditions”, *IEEE Transactions on Applied Superconductivity* Vol. 3 No. 1 (1993). DOI: 10.1109/77.233691

[109] M. N. Wilson, *Superconducting magnets*. Oxford University Press (1983). ISBN: 978-0198548102

[110] A. Devred, L. Bacquart, P. Bredy, C. E. Bruzek, Y. Laumond et al., "Interstrand resistance measurements on Nb_3Sn Rutherford-type cables", *IEEE Transactions on Applied Superconductivity* Vol. 9 No. 2 (1999). DOI: 10.1109/77.783397

[111] J. Fleiter, A. Ballarino, L. Bottura, W. Goldacker and A. Kario, "Characterization of Roebel Cables for Potential Use in High-Field Magnets", *IEEE Transactions on Applied Superconductivity* Vol. 25 No. 3 (2015). DOI: 10.1109/TASC.2014.2367162

[112] S. Schuller, W. Goldacker, A. Kling, L. Krempasky and C. Schmidt, "Ac-loss measurement of a DyBCO-Roebel assembled coated conductor cable (RACC)", *Physica C: Superconductivity and its Applications* Vol. 463 (2007). DOI: 10.1016/j.physc.2007.01.063

[113] C. Schmidt, A. Frank, W. Goldacker, A. Kling and S. Terzieva, "Progress in assembling coated conductor cables by the Roebel technique (RACC)", *Physica C: Superconductivity and its Applications* Vol. 469 No. 15 (2009). DOI: 10.1016/j.physc.2009.05.050

[114] L. S. Lakshmi, N. J. Long, R. A. Badcock, M. P. Staines, Z. Jiang et al., "Magnetic and Transport AC Losses in HTS Roebel Cable", *IEEE Transactions on Applied Superconductivity* Vol. 21 No. 3 (2011). DOI: 10.1109/TASC.2010.2088369

[115] L. S. Lakshmi, M. P. Staines, R. A. Badcock, N. J. Long, M. Majoros et al., "Frequency dependence of magnetic ac loss in a Roebel cable made of YBCO on a Ni-W substrate", *Superconductor Science and Technology* Vol. 23 No. 8 (2010). DOI: 10.1088/0953-2048/23/8/085009

[116] E. B. Rosa, "The self and mutual inductances of linear conductors", *Bulletin of the Bureau of Standards* Vol. 4 No. 2 (1908)

[117] G. Ries and S. Takács, "Coupling losses in finite length of superconducting cables and in long cables partially in magnetic field", *IEEE Transactions on Magnetics* Vol. 17 No. 5 (1981). DOI: 10.1109/TMAG.1981.1061354

[118] E. H. Brandt, "Thin superconductors in a perpendicular magnetic ac field: General formulation and strip geometry", *Physical Review B* Vol. 49 No. 13 (1994). DOI: 10.1103/PhysRevB.49.9024

[119] D. J. Griffiths, *Introduction to Electrodynamics*, 3rd ed. Pearson Benjamin Cummings (2008). ISBN: 0-13-919960-8

[120] J. Ekin, *Experimental Techniques for Low-Temperature Measurements: Cryostat Design, Material Properties and Superconductor Critical-Current Testing*. Oxford university press (2006). ISBN: 978-0-19-857054-7

[121] M. R. Halse, "AC face field losses in a type II superconductor", *Journal of Physics D: Applied Physics* Vol. 3 No. 5 (1970). DOI: 10.1088/0022-3727/3/5/310

[122] E. H. Brandt and M. Indenbom, "Type-II-superconductor strip with current in a perpendicular magnetic field", *Physical Review B* Vol. 48 No. 17 (1993). DOI: 10.1103/PhysRevB.48.12893

[123] M. Sander and F. Grilli, "FEM-calculations on the frequency dependence of hysteretic losses in coated conductors", *Journal of Physics: Conference Series* Vol. 234 No. 2 (2010). DOI: 10.1088/1742-6596/234/2/022030

[124] S. Terzieva, M. Vojenčiak, E. Pardo, F. Grilli, A. Drechsler et al., "Transport and magnetization ac losses of ROEBEL assembled coated conductor cables: measurements and calculations", *Superconductor Science and Technology* Vol. 23 No. 1 (2010). DOI: 10.1088/0953-2048/23/1/014023

[125] V. M. R. Zermeño, K. Habelok, M. Stępień and F. Grilli, "A parameter-free method to extract the superconductor's $J_c(B,\theta)$ field-dependence from in-field current-voltage characteristics of high temperature superconductor tapes", *Superconductor Science and Technology* Vol. 30 No. 3 (2017). DOI: 10.1088/1361-6668/30/3/034001

[126] J. Šouc, F. Gömöry and M. Vojenčiak, "Calibration free method for measurement of the AC magnetization loss", *Superconductor Science and Technology* Vol. 18 No. 5 (2005). DOI: 10.1088/0953-2048/18/5/003

[127] M. Polak, J. Kvitkovic, P. Mozola, E. Usak, P. Barnes et al., "Frequency dependence of hysteresis loss in YBCO tapes", *Superconductor Science and Technology* Vol. 20 No. 9 (2007). DOI: 10.1088/0953-2048/20/9/S26

[128] S. Hahn, D. K. Park, J. Bascunan and Y. Iwasa, "HTS Pancake Coils Without Turn-to-Turn Insulation", *IEEE Transactions on Applied Superconductivity* Vol. 21 No. 3 (2011). DOI: 10.1109/TASC.2010.2093492

[129] D. G. Yang, J. B. Song, Y. H. Choi, M. C. Ahn, J. B. Na et al., "Quench and Recovery Characteristics of the Zr-Doped (Gd,Y) BCO Coated Conductor Pancake Coils Insulated With Copper and Kapton Tapes", *IEEE Transactions on Applied Superconductivity* Vol. 21 No. 3 (2011). DOI: 10.1109/TASC.2010.2090323

[130] H. J. Shin, K. L. Kim, Y. H. Choi, D. Yang, Y. G. Kim et al., "A Study on Normal Zone Propagation Behavior of Partially Insulated GdBCO Coil", *IEEE Transactions on Applied Superconductivity* Vol. 25 No. 3 (2015). DOI: 10.1109/TASC.2014.2363031

[131] Y. Wang, W. K. Chan and J. Schwartz, "Self-protection mechanisms in no-insulation (RE)$Ba_2Cu_3O_x$ high temperature superconductor pancake coils", *Superconductor Science and Technology* Vol. 29 No. 4 (2016). DOI: 10.1088/0953-2048/29/4/045007

[132] A. K. Gosh, W. B. Sampson and M. N. Wilson, "Minimum quench energies of Rutherford cables and single wires", *IEEE Transactions on Applied Superconductivity* Vol. 7 No. 2 (1997). DOI: 10.1109/77.614663

[133] J. van Nugteren, *High Temperature Superconducting Accelerator Magnets*. PhD thesis, University of Twente (2016). ISBN: 978-94-028-0403-4

[134] A. Ishiyama, M. Yanai, T. Morisaki, H. Ueda, Y. Shiohara et al., "Normal transition and propagation Characteristics of YBCO tape", *IEEE Transactions on Applied Superconductivity* Vol. 15 No. 2 (2005). DOI: 10.1109/TASC.2005.849225

[135] L. Bottura, M. Breschi and M. Fabbri, "An analytical benchmark for the calculation of current distribution in superconducting cables", *Cryogenics* Vol. 43 No. 3 (2003). DOI: 10.1016/S0011-2275(03)00041-9

[136] H. Bai, W. D. Markiewicz, J. Lu and H. W. Weijers, "Thermal Conductivity Test of YBCO Coated Conductor Tape Stacks Interleaved With Insulated Stainless Steel Tapes", *IEEE Transactions on Applied Superconductivity* Vol. 23 No. 3 (2013). DOI: 10.1109/TASC.2012.2229774

[137] N. Bagrets, W. Goldacker, A. Jung and K. P. Weiss, "Thermal Properties of ReBCO Copper Stabilized Superconducting Tapes", *IEEE Transactions on Applied Superconductivity* Vol. 23 No. 3 (2013). DOI: 10.1109/TASC.2012.2233268

[138] M. Bonura and C. Senatore, "Transverse Thermal Conductivity of REBCO Coated Conductors", *IEEE Transactions on Applied Superconductivity* Vol. 25 No. 3 (2015). DOI: 10.1109/TASC.2014.2367163

[139] K. Morton and D. Mayers, *Numerical Solution of Partial Differential Equations*. Cambridge University Press (2005). ISBN: 978-0-521-60793-3

[140] Y. Liu, J. Ou, F. Schreiner, M. Lao, M. Noe et al., "Design of a Superconducting DC Demonstrator for Wind Generators", *IEEE Transactions on Energy Conversion* (2018). DOI: 10.1109/TEC.2018.2846721

[141] M. Bonura and C. Senatore, "High-field thermal transport properties of REBCO coated conductors", *Superconductor Science and Technology* Vol. 28 No. 2 (2015). DOI: 10.1088/0953-2048/28/2/025001

[142] J. Lu, E. S. Choi and H. D. Zhou, “Physical properties of Hastelloy C-276 at cryogenic temperatures”, *Journal of Applied Physics* Vol. 103 No. 6 (2008). DOI: 10.1063/1.2899058

[143] N. Bagrets, W. Goldacker, S. I. Schlachter, C. Barth and K.-P. Weiss, “Thermal properties of 2G coated conductor cable materials”, *Cryogenics* Vol. 61 (2014). DOI: 10.1016/j.cryogenics.2014.01.015

[144] C. Kittel, *Introduction to Solid State Physics*, 8th ed. Wiley (2004). ISBN: 978-0-471-41526-8

[145] J. Pelegrin, E. Martinez, L. A. Angurel, Y. Y. Xie and V. Selvamanickam, “Numerical and Experimental Analysis of Normal Zone Propagation on 2G HTS Wires”, *IEEE Transactions on Applied Superconductivity* Vol. 21 No. 3 (2011). DOI: 10.1109/TASC.2010.2084982

[146] K. Seo, M. Morita, S. Nakamura, T. Yamada and Y. Jizo, “Minimum quench energy measurement for superconducting wires”, *IEEE Transactions on Magnetics* Vol. 32 No. 4 (1996). DOI: 10.1109/20.511529

[147] N. Amemiya, H. Yonekawa, T. Ogitsu, E. Kobayashi, K. Sasaki et al., “Influence of current re-distribution on minimum quench energy of superconducting triplex cable against local disturbance”, *Cryogenics* Vol. 38 No. 5 (1998). DOI: 10.1016/S0011-2275(98)00006-X

[148] A. Ghosh, W. Sampson, S. Kim, D. Leroy, L. Oberli et al., “Stability measurements on cored cables in normal and superfluid helium”, *Physica C: Superconductivity and its Applications* Vol. 310 No. 1 (1998). DOI: doi.org/10.1016/S0921-4534(98)00484-5

[149] N. Amemiya, H. Yonekawa, E. Kobayashi, T. Ogitsu, K. Sasaki et al., “Current redistribution and stability of superconducting triplex cable without electrical insulation carrying non-uniform current”, *Cryogenics* Vol. 43 No. 3 (2003). DOI: 10.1016/S0011-2275(03)00042-0

[150] H. Dolmetsch, W. Klein, D. Holznagel, T. Odenwald and E. Keller, *Metalltechnik Fachbildung: Der Werkzeugbau*. Verlag Europe-Lehrmittel - Nourney, Vollmer GmbH Co. KG (2011). ISBN: 978-3808512005

[151] A. Badel, A. Ballarino, C. Barth, L. Bottura, M. M. J. Dhallé et al., “Advances in the Development of a 10-kA Class REBCO Cable for the EuCARD2 Demonstrator Magnet”, *IEEE Transactions on Applied Superconductivity* Vol. 26 No. 3 (2016). DOI: 10.1109/TASC.2016.2548938

[152] N. Long, E. Talantsev, S. Chong, R. Badcock and K. Bouloukakis, "Effect of random position shifts of strands on the transverse mechanical properties of REBCO Roebel cables", *Presented at the European Conference on Applied Superconductivity, Lyon, France* (2015)

[153] A. Kario, S. Otten, A. Kling, U. Mirasch, A. Drechsler et al., "Investigation of the Roebel cable geometry and current homogeneity over lengths relevant to accelerator-type demonstrator magnet", *Presented at the 25th International Conference on Magnet Technology (MT25), Amsterdam, The Netherlands* (2017). URL: https://indico.cern.ch/event/445667/contributions/2561930/

Karlsruher Schriftenreihe zur Supraleitung

Karlsruher Institut für Technologie (KIT) | ISSN 1869-1765

Herausgeber: Prof. Dr.-Ing. M. Noe, Prof. Dr. rer. nat. M. Siegel

Band 001 **Christian Schacherer**
Theoretische und experimentelle Untersuchungen zur Entwicklung supraleitender resistiver Strombegrenzer. 2009
ISBN 978-3-86644-412-6

Band 002 **Alexander Winkler**
Transient behaviour of ITER poloidal field coils. 2011
ISBN 978-3-86644-595-6

Band 003 **André Berger**
Entwicklung supraleitender, strombegrenzender Transformatoren. 2011
ISBN 978-3-86644-637-3

Band 004 **Christoph Kaiser**
High quality Nb/Al-AlOx/Nb Josephson junctions. Technological development and macroscopic quantum experiments. 2011
ISBN 978-3-86644-651-9

Band 005 **Gerd Hammer**
Untersuchung der Eigenschaften von planaren Mikrowellen-resonatoren für Kinetic-Inductance Detektoren bei 4,2 K. 2011
ISBN 978-3-86644-715-8

Band 006 **Olaf Mäder**
Simulationen und Experimente zum Stabilitätsverhalten von HTSL-Bandleitern. 2012
ISBN 978-3-86644-868-1

Die Bände sind unter www.ksp.kit.edu als PDF frei verfügbar oder als Druckausgabe bestellbar.

Band 007 **Christian Barth**
High Temperature Superconductor Cable Concepts
for Fusion Magnets. 2013
ISBN 978-3-7315-0065-0

Band 008 **Axel Stockhausen**
Optimization of Hot-Electron Bolometers for THz Radiation. 2013
ISBN 978-3-7315-0066-7

Band 009 **Petra Thoma**
Ultra-fast $YBa_2Cu_3O_{7-x}$ direct detectors for the THz
frequency range. 2013
ISBN 978-3-7315-0070-4

Band 010 **Dagmar Henrich**
Influence of Material and Geometry on the Performance
of Superconducting Nanowire Single-Photon Detectors. 2013
ISBN 978-3-7315-0092-6

Band 011 **Alexander Scheuring**
Ultrabreitbandige Strahlungseinkopplung in THz-Detektoren. 2013
ISBN 978-3-7315-0102-2

Band 012 **Markus Rösch**
Development of lumped element kinetic inductance detectors
for mm-wave astronomy at the IRAM 30 m telescope. 2013
ISBN 978-3-7315-0110-7

Band 013 **Johannes Maximilian Meckbach**
Superconducting Multilayer Technology for Josephson
Devices. 2013
ISBN 978-3-7315-0122-0

Band 014 **Enrico Rizzo**
Simulations for the optimization of High Temperature
Superconductor current leads for nuclear fusion applications. 2014
ISBN 978-3-7315-0132-9

Band 015 **Philipp Krüger**
Optimisation of hysteretic losses in high-temperature superconducting wires. 2014
ISBN 978-3-7315-0185-5

Band 016 **Matthias Hofherr**
Real-time imaging systems for superconducting nanowire single-photon detector arrays. 2014
ISBN 978-3-7315-0229-6

Band 017 **Oliver Näckel**
Development of an Air Coil Superconducting Fault Current Limiter. 2016
ISBN 978-3-7315-0526-6

Band 018 **Christoph M. Bayer**
Characterization of High Temperature Superconductor Cables for Magnet Toroidal Field Coils of the DEMO Fusion Power Plant. 2017
ISBN 978-3-7315-0605-8

Band 019 **Shengnan Zou**
Magnetization of High Temperature Superconducting Trapped-Field Magnets. 2017
ISBN 978-3-7315-0715-4

Band 020 **Ilya Charaev**
Improving the Spectral Bandwidth of Superconducting Nanowire Single-Photon Detectors (SNSPDs). 2018
ISBN 978-3-7315-0745-1

Band 021 **Juliane Raasch**
Electrical-field sensitive $YBa_2Cu_3O_{7-x}$ detectors for real-time monitoring of picosecond THz pulses. 2019
ISBN 978-3-7315-0786-4

Band 022 **Yingzhen Liu**
Design of a superconducting DC wind generator. 2019
ISBN 978-3-7315-0796-3

Die Bände sind unter www.ksp.kit.edu als PDF frei verfügbar oder als Druckausgabe bestellbar.

Band 023 **Sebastian Hellmann**
Research and Technology Development on Superconducting Current Limiting Transformers. 2019
ISBN 978-3-7315-0804-5

Band 024 **Simon J. Otten**
Characterisation of REBCO Roebel cables. 2019
ISBN 978-3-7315-0904-2